Computer Algebra

Systems and algorithms
for algebraic Computation

The diagram "Constructeur Universel d'Equations" (Universal Constructor of Equations) on the front cover is taken from the third volume of illustrations for Diderot's Encyclopédie. The construction and use of this machine to find the roots of an equation of arbitary degree is described by d'Alembert (Diderot's principal collaborator, and the mathematical editor of the encyclopedia) in volume twelve.

Diderot's Encyclopédie (1751–1772) was the first French encyclopedia, the work of a great community of French scientists and philosophers directed by Diderot and d'Alembert. It served as the model for most subsequent encyclopedias.

Computer Algebra

Systems and algorithms
for algebraic computation

J. H. Davenport

School of Mathematical Sciences
University of Bath
Bath

Y. Siret

Centre Interuniversitaire
de Calcul de Grenoble
Grenoble

E. Tournier

Université Scientifique, Technologique
et Médicale de Grenoble
Grenoble

Translated from the French by
A. Davenport and J. H. Davenport

1988

Academic Press
Harcourt Brace Jovanovich, Publishers
London   San Diego   New York   Boston
Sydney   Tokyo   Toronto

ACADEMIC PRESS LIMITED
24–28 Oval Road, London NW1 7DX

United States Edition published by
ACADEMIC PRESS INC.
San Diego, CA 92101

British Library Cataloguing in Publication Data

Is Available

ISBN 0-12-204230-1

Originally published in French
by Masson, Paris

Printed by St Edmundsbury Press Limited,
Bury St Edmunds, Suffolk

Preface

An example

The example which follows is a work session on the computer, using the software MACSYMA. The user's commands are entered in the lines beginning with (ci). Here we are dealing with a limited (or Taylor series) expansion, calculating an integral and checking that the derivative in the result is indeed the initial function.

```
(c2) taylor( sinh(sin(x))-sin(sinh(x))   ,x ,0 , 15);
                  7    11             15
                 x    x         5699 x
(d2)/T/          -- - ---- - ---------- + . . .
                 45   1575    1277025750


(c3) primitive:integrate( 1/(x**2 + 1) ** 4 ,x);
                                   5       3
              5 atan(x)       15 x  + 40 x  + 33 x
(d3)          ---------  + --------------------------
                 16           6        4        2
                          48 x  + 144 x  + 144 x  + 48


(c4) derivative:diff(primitive, x);
                 4        2
           75 x  + 120 x  + 33
(d4)    ---------------------------
            6        4        2
        48 x  + 144 x  + 144 x  + 48

         5        3              5        3
    (15 x  + 40 x  + 33 x) (288 x  + 576 x  + 288 x)          5
  - ------------------------------------------------- + -----------
            6        4        2        2                      2
        (48 x  + 144 x  + 144 x  + 48)                   16 (x  + 1)
```

```
(c5) factor(ratsimp(derivative));
```

$$(d5) \qquad \frac{1}{(x^2 + 1)^4}$$

All these calculations can, in theory, be done by a first year student, but they are quite laborious and it is difficult to do them with pencil and paper without making a mistake.

It is not sufficiently well-known that this type of calculation is as much within the range of computers as is numerical calculation. The purpose of this book is to demonstrate the existing possibilities, to show how to use them, and to indicate the principles on which systems of non numerical calculation are based, to show the difficulties which the designers of these systems had to solve — difficulties with which the user is soon faced.

But before we try to describe how the authors hope to carry out this ambitious design, we should first try to define the subject better.

Scientific calculation and algebraic calculation

From the start of electronic calculation, one of the main uses of computers has been for numerical calculation. Very soon, applications to management began to dominate the scene, as far as the volume of calculation assigned to them is concerned. Nevertheless, scientific applications are still the most prestigious, especially if we look at the performance required of the computer: the most powerful computers are usually reserved for scientific calculation.

This concept of "scientific calculation" conceals an ambiguity, which it is important to note: before computers appeared on the scene, a calculation usually consisted of a mixture of numerical calculation and what we shall call "algebraic calculation", that is calculation by mathematical formulae. The only example of purely numerical calculation seems to have been the feats of calculating prodigies such as Inaudi: the authors of tables, especially of logarithms, did indeed carry out enormous numerical calculations, but these were preceded by a restatement of the algebraic formulae and methods which were essential if the work was to be within the bounds of what is humanly possible. For example the famous large calculations of the 19th century include a large proportion of formula manipulation. The best known is certainly Le Verrier's calculation of the orbit of Neptune, which started from the disturbances of the orbit of Uranus and which led to the discovery of Neptune. The most impressive calculation with pencil and paper is also in the field of astronomy: Delaunay took 10 years to calculate the orbit of the moon, and another 10 years to check it. The result is not numerical, because it consists for the most part of a formula which by itself occupies all the 128 pages of Chapter 4 of his book.

The ambiguity mentioned above is the following: when computers came on the scene, numerical calculation was made very much easier and it became commonplace to do enormous calculations, which in some cases made it possible to avoid laborious algebraic manipulations. The result was that, for the public at large and even for most scientists, numerical calculation and scientific calculation have become synonymous. When the title or list of subjects of a congress or conference includes the words "scientific calculation", we can generally be sure that it is a question of numerical calculation only, even if Computer Algebra is also in the list of subjects.

However, numerical calculation does not rule out algebraic calculation: writing the most trivial numerical program requires a restatement of the formulae on which the algorithm is based. And, the power of computers does not solve everything: calculating the development of the atmosphere with the precision needed by meteorologists for a forecast 48 hours ahead would require much more than 48 hours on the most powerful computer at present available (CRAY). Multiplying their power by 10 would make the situation only marginally better for, at the same time, the theory would be being improved and the mesh-size would have to be divided by 2 (to distinguish for example between the weather in Paris and that in Orléans), and this would multiply by 8 the number of numerical calculations needed.

Computer Algebra

Thus algebraic calculation has not lost its relevance. However, it is most frequently done with pencil and paper, even though the first software intended to automate it is already quite old [Kahrimanian, 1953; Nolan, 1953]. It was very quickly seen that such software for helping algebraic calculation would have to be a complete system, which included a method with a very special structure for representing non numerical data, a language making it possible to manipulate them, and a library of effective functions for carrying out the necessary basic algebraic operations.

And so there appeared systems which run very well, and the most widely used of these are described or referred to in this book.

Writing, developing and even using these systems gradually merged into an autonomous scientific discipline, obviously based on computer science. But the objectives are in the field of artificial intelligence, even if the methods are moving further and further away from it. Moreover, the algorithms used bring into play less and less elementary mathematical tools. And so this discipline seems to form a frontier between several fields, and this adds both to its richness and, from the research aspect, to its difficulty.

The name of this discipline has long hesitated between "symbolic and algebraic calculation", "symbolic and algebraic manipulations", and finally settled down as "Computer Algebra" in English, and "Calcul Formel", abbreviated to CALSYF, in French. At the same time societies were formed to bring together the research workers and the users of this discipline: the world-wide organisation is SIGSAM[1] group of the ACM[2] which organises the congresses SYMSAC and EUROSAM and publishes the bulletin SIGSAM. The European group is called SAME[3] and organises congresses called EUROCAM, EUROSAM,... the proceedings of which are referred to in the bibliography. For French research workers there is the body of the CNRS (Centre national de la recherche scientifique), the GRECO[4] of Computer Algebra. And since 1985 a specialist review the *Journal of Symbolic Computation* has been published.

Computer Algebra systems

There are very many computer algebra systems; but scarcely more than ten are up-to-date, general and fairly widely available. In this book the authors have chosen four as being especially representative:
- MACSYMA, the most developed, but unfortunately available on only very few kinds of computers.
- REDUCE, the most widely available of all the large systems.
- muMATH, the system most easily available to micro-computers, which means that it is very widely used but at the same time means that it is mainly used by beginners.
- SCRATCHPAD, a very new system, with extremely restricted availability, but which, because of its completely different structure, can overcome some limitations the other systems share, and can be the prototype for the next generation of Computer Algebra systems.

With the exception of SCRATCHPAD, the 3 other systems mentioned here and most of those not mentioned are very user friendly; admittedly the superficial syntax of the language is not the same; admittedly, the library of available functions varies from some dozens to more than a thousand; admittedly, the internal structure of the system varies considerably; but they all share the following properties:
- The programming is mainly interactive: the user does not, in theory, know either the form or the size of his results and must therefore be able to intervene at any time.
- Most of the data worked on are mathematical expressions, of which at least the external representation is the one everyone is accustomed to.
- The language used is "ALGOL-like".

---

[1] Special Interest Group in Symbolic and Algebraic Manipulations.
[2] Association for Computing Machinery.
[3] Symbolic and Algebraic Manipulations in Europe.
[4] Groupe de REcherches COordonnées.

- The implementation language is often LISP; in any case, the data are list structures and tree structures, and memory management is dynamic with automatic recovery of the available space.

Thus, once one of the Computer Algebra systems has been mastered, changing to a different one does not usually present the programmer with many problems, with far fewer in any case than does changing the system used at the same time as changing computer.

That is why the authors thought it better to write a general introduction to Computer Algebra than a manual for a particular system.

And so, the introductory chapter "How to use a Computer Algebra system" is written in MACSYMA, the commands and the results are extremely readable and easily understood. But REDUCE has been chosen for the detailed description of a system, because it is more widely available. The other chapters do not usually refer to a particular system.

Accessibility of Computer Algebra systems

Computer Algebra systems can generally be used only on fairly large machines whose operating system works in virtual memory: to work without too much difficulty, a mega-byte of main memory seems to be the smallest amount suitable. The most economical system is obviously muMATH; the most costly in work memory is SCRATCHPAD for which 8 mega-bytes of work space are necessary at present. The one which needs most disk space is probably MACSYMA whose compiled code is getting on for mega-bytes, to which must be added several mega-bytes of line documentation.

At present the availability of the systems is as follows: MACSYMA runs only on VAX, MULTICS (this system, fairly wide-spread in France, is unfortunately doomed to disappear) and some personal work-stations (Symbolics, Sun). REDUCE can be used on almost all the main models of large machines or of mini-computers, even though in some cases only the old versions are available (Cyber, IBM series 360, 370,..., MULTICS, all the UNIX systems,...). muMATH can be used on almost every micro-computer. As for SCRATCHPAD, it only works on IBM's VM/CMS system, and for the present only a very small number of implementation sites per country are planned.

Using Computer Algebra systems

For a beginner, the Computer Algebra languages are some of the simplest to use. In fact, at first, he only needs to know some functions which allow him to rewrite the problem in question in a form very similar to its mathematical form. Even if the rewriting is clumsy or incorrect, the interactivity means that after some playing around he can quickly find results unobtainable with pencil and paper. And, for many applications, that is sufficient.

In programming language such as FORTRAN the syntactical subtleties require a long apprenticeship whereas the work principles of the compiler can be totally ignored, but here, the user must very quickly grasp "how that works", especially how the data are represented and managed.

In fact, although it is usually difficult to foretell the time a calculation will take and what its size will be, a knowledge of the working principles can give an idea as to their order of magnitude and, if need be, optimise them. These estimates are in fact essential: for most algebraic calculations, the results are quasi-instantaneous, and all goes well. But, if this is not so, the increases in the time and memory space required are usually exponential. So the feasibility of a given calculation is not always obvious, and it is stupid to commit large resources when a failure can be predicted.

Moreover, an intelligent modelling of the problem and a program adapted to the structure of the data may make an otherwise impossible problem easy. Thus, in Chapter 1 two programs for calculating the largest coefficient of a polynomial are given; the text of the one which seems most natural for someone used to FORTRAN is considerably more complicated than the other, but above all it needs 10 times more time for a polynomial of degree 30 ; if we move on to a polynomial of degree 1000, about 30 times greater, the running time of the second program is in theory multiplied by 30, whereas that of the first is multiplied by a factor of the order of 1000, which gives a running time ratio of 300 between the two programs, and real times of about one minute and five hours.

Acquiring an efficient programming style and an ability to foresee the size of a calculation is therefore much more important here than in numerical calculation where the increase is generally linear. Unfortunately, this is for many a matter of experience, which it is hard to pass on by way of a manual. However, familiarity makes it easier to acquire this experience; and that is why such a large part of the book is devoted to describing the mathematical working of Computer Algebra systems.

Plan of the book

The book can be divided into four sections.

The first (Chapter 1) is an introduction to Computer Algebra with the help of annotated examples, most of which are written in MACSYMA. They can usually be understood on a first reading. Nevertheless, it is a very useful exercise to program these examples or similar ones in MAC-SYMA or in some other system. This obviously requires a knowledge of the corresponding users' manual.

The second part, which consists of Chapters 2, 3, and 4, describes the working of Computer Algebra systems. Besides being interesting in itself, the description of the problems which had to be solved and of the principles behind the chosen solutions is essential for helping the user to get round the problems of "combinatorial explosion" with which he will be confronted.

Chapter 5 makes up the third part and gives two Computer Algebra problems, still at the research stage even though the softwares described are beginning to be widely used. They are very good examples, which require all the power and all the possibilities of Computer Algebra systems to solve difficult natural problems; they are good illustrations both of the possibilities and of the difficulties inherent in this discipline.

Finally, a detailed presentation of REDUCE, as an appendix, enables this book to serve as a manual, at least for the most widely used of these systems.

There is also a detailed bibliography, which includes many more references than appear explicitly in the text, so that the reader has access to the methods, results and algorithms for which there was no room in the book.

In conclusion, let us hope that this book, which till now has no parallel anywhere in the world, will help to increase the use of this tool, so powerful and so little known: Computer Algebra.

Daniel Lazard
Head of Computer Algebra GRECO
Professor at the P. and M. Curie University (Paris VI)

ACKNOWLEDGEMENTS

The authors would like to thank all those who read the various drafts of this book, in particular Mrs. H. Davenport, J. Della Dora, D. Duval, M. Giusti, D. Lazard and J.C. Smith.

This book was typeset with the TEX system. The author of TEX, D.E. Knuth, is responsible for the typographical excellence of the book, but obviously not for the errors. The presentation owes much to the TEX experts: C. Goutorbe at the C.I.C.G.* and C.E. Thompson of the University of Cambridge.

---

* The French original was prepared on the hardware of the C.I.C.G. (Inter-University Computer Centre of Grenoble).

Foreword

It is now over thirty years since the computer was first used to perform algebraic calculations. The resulting programs could differentiate very simple expressions, a far cry from the impressive array of sophisticated calculations possible with today's algebraic computation programs. Until recently, however, such programs have only been accessible to those with access to large main-frame computers. This has limited both the use and the appreciation of algebraic computation techniques by the majority of potential users. Fortunately, the personal computer revolution is changing this; readily available machines are now appearing that can run existing algebra programs with impressive ease. As a result, interest in computer algebra is growing at a rapid pace.

Although programs exist for performing the esoteric algebraic computations of interest to professional mathematicians, in this book we are mainly concerned with the constructive methods used by physical scientists and engineers. These include such mundane things as polynomial and series manipulation, as well as more sophisticated techniques like analytic integration, polynomial factorization and the analytic solution of differential equations. In addition, a growing number of scientists are finding that these programs are also useful for generating numerical code from the resulting expressions.

Algebraic computation programs have already been applied to a large number of different areas in science and engineering. The most extensive use has occurred in the fields where the algebraic calculations necessary are extremely tedious and time consuming, such as general relativity, celestial mechanics and quantum chromodynamics. Before the advent of computer algebra, such hand calculations took many months to complete and were error prone. Personal workstations can now perform much larger calculations without error in a matter of minutes.

Given the potential of computer algebra in scientific problem solving, it is clear that a text book is needed to acquaint users with the available techniques. There are books concerned with the use of specific algebraic computation programs, as well as those oriented towards computer scientists who study the field. However, this is the first general text to appear that provides the practising scientist with the information needed to make optimal use of the available programs. The appearance of an English version of the original French text also makes it accessible to a wider class of readers than was previously possible.

In addition to being a general text on the subject, this book also includes an annex describing the use of one particular algebra system, namely REDUCE, a system I first designed in the late 1960's. The version described here has undergone many changes and extensions since those days, and is now in use by thousands of scientists and engineers throughout the world on machines ranging in power from the IBM PC to the Cray X-MP. I am sure that they will welcome this useful exposition of the system.

Anthony C. Hearn
The RAND Corporation
December 1987

ACKNOWLEDGEMENTS

The translators would like to thank J.P. Fitch and J.C. Smith for their helpful comments on this translation, as well as the reader R. Cutler at Academic Press. This translation was also prepared with TeX, and C.E. Thompson again provided much useful advice. The data transfers involved in preparing this translation would have been impossible without the helpful advice and support of the University of Cambridge Computing Service. For the re-printing, R.J. Bradford and M.A.H. MacCallum signalled a few mis-prints, which have been corrected.

Contents

1. How to use a Computer Algebra system

1.1 INTRODUCTION

The idea of doing algebraic calculation on computers is not new. From 1960 on, numerous programs have appeared on the market which are intended to show that in the scientific field one can go beyond the purely numerical area usually attributed to computers. The language LISP dates from this period and opened the way to the first spectacular demonstrations of the following possibilities: formal integration and proofs of theorems. In the same vein, some people very quickly saw that they could ask the machine to carry out algebraic operations which, though tedious, were useful for later numerical calculation: the expansion of polynomial expressions, formal differentiation etc.... The appearance of time-sharing systems contributed a great deal to the generalisation of programs of algebraic calculation, and they have gradually become genuine sub-systems which can be used in a language close to the usual algorithmic languages. But as the problem is much bigger than numerical calculation or management there has never been any standardisation, as there has been for those languages. That is one of the reasons why these possibilities have so long been ignored by the scientific and industrial world. For a long time too the use of these systems has been limited to a certain type of machine and obviously this limitation has not helped to spread knowledge of them. However, recently things have been changing: the language LISP is finding favour again and its numerous dialects and variations have not prevented a system such as REDUCE from being available on a great number of computers and work-stations. To a slightly lesser extent, MACSYMA has been adapted for commercial use on a certain number of computers and is no longer considered to be an inaccessible system. More recently there have appeared some systems for micro-computers which, albeit limited, at least give us some feeling for

1

the subject: muMATH is one of these. It is very remarkable that the main existing systems: REDUCE, MACSYMA, muMATH and the promising system SCRATCHPAD are written in LISP. A knowledge of this language will certainly help to make mechanical algebraic calculation better understood and used.

1.2 FEATURES OF COMPUTER ALGEBRA SYSTEMS

The most intuitive, although somewhat restrictive, approach to Computer Algebra systems is to say that they are made for the manipulation of everyday scientific and engineering formulae. A mathematical formula which is described in one of the usual languages (FORTRAN, PASCAL, BASIC,...) can only be evaluated numerically, once the variables and parameters have themselves been given numerical values. In a language which allows of algebraic manipulations, the same formula can also be evaluated numerically, but above all it can be the object of formal transformations: differentiation, development in series, various expansions, even integration.

For example, one can, with just one command, ask for the decomposition into partial fractions of

$$\frac{x^2 - 5}{x(x-1)^4}$$

which is

$$\frac{-5}{x} + \frac{5}{x-1} - \frac{5}{(x-1)\,2} + \frac{6}{(x-1)\,3} - \frac{4}{(x-1)\,4}$$

and fairly easy to do by hand. But it is nice to do by machine the decomposition of

$$\frac{x+a}{x(x-b)(x^2+c)}$$

which is a rather more complicated form:

$$-\frac{a}{bcx} + \frac{a+b}{(bc+b^3)(x-b)} - \frac{(c-ab)x + (b+a)c}{(c^2+b^2c)(x^2+c)}.$$

As a general rule, Computer Algebra systems are drawn up to meet the two following requirements:

- To provide a set of basic pre-programmed commands which will hand over to the machine the wearisome calculations which occur in a process run completely by the user: this is the role of an advanced desk-top machine.

- To offer a programming language which lets us define higher-level commands or procedures to enlarge the original set of commands.

We see that there is no fundamental conflict with the aims of languages such as LISP or APL: the same conversational aspect, the same possibilities for defining procedures or functions and thus the same possibilities for adding to the existing libraries on a given subject.

Naturally, when the algorithm is suitable, it is perfectly possible to exploit algebraic calculation as a batch job. That depends only on the computing environment and it is up to the user to see that all the data are defined at the time of execution.

1.3 SYNTAX OF THE ASSOCIATED LANGUAGES

Although there are some differences between the systems, syntax is not the main problem of Computer Algebra. In general a few hours and a little practice are perfectly adequate. The syntax can be said to be largely that of PASCAL. Of course there is an assignment instruction, the idea of calling functions (commands), a fairly rich set of control structures (if, do, while, repeat etc....), possibilities for declaring procedures .... In brief, all the arsenal of programming languages required for writing algorithms.

1.4 AREAS COVERED BY EXISTING SYSTEMS

If it is to be useful and to keep the user's interest, a Computer Algebra system must cover by means of a rich set of commands those areas which immediately come to mind when one puts aside the limitations of hardware, and of traditional programming languages. The following tools are part of the necessary equipment:

- Operations on integers, on rational, real and complex numbers *with unlimited accuracy*. Chapter 2 shows the importance of this.
- Operations on polynomials in one or more variables and on rational fractions. In short, the obvious rational operations, calculating the g.c.d., factorising over the integers.
- Calculations on matrices with numerical and/or symbolic elements.
- Simple analysis: differentiation, expansion in series, Padé approximants etc.
- Manipulation of formulae: various substitutions, selection of coefficients and of parts of formulae, numerical evaluation, pattern recognition, controlled simplifications, ...

Starting out from this common base, the systems may offer possibilities in specific areas, possibilities which to some extent characterise their degree of development. For example:

- Solution of equations.
- Formal integration.
- Calculation of limits.
- Tensor calculus.

In addition, users have an opportunity to add to the function library by passing on work which is sufficiently general to interest a group of people. A typical example is the SHARE library of MACSYMA.

1.5 COMPUTER ALGEBRA BY EXAMPLE

In what follows we have chosen to expound, using the MACSYMA system, a series of examples from some of the areas mentioned. Most of the examples can be repeated using other systems, especially REDUCE which is described in the annex. It is obvious that we are not trying to replace the MACSYMA programmers' manual and the reader will at times have to be satisfied with brief explanations given in the commentary. The examples are printed as they were given to the machine. The typography of the results has not been changed. The lines typed by the user are labelled (c_i) and finish with ";" or "$". The latter symbol tells MACSYMA not to print the result. So time is saved if printing the result is not essential. The result of a command is labelled (d_i). The symbol % used in a command is the value associated with the previous command. The constants π, e, i are written %pi, %a, %i. The assignment symbol is the *colon*.

1.5.1 Simple operations on numbers

```
(c1) n0 : 123456789123456789 * 123456789 + 987654321123456789;

(d1)                          1524157975308642087 3647310

(c2) n1 : 100! ;

(d2) 93326215443944152681699238856266700490715968264381621

     4685929638952175999932299156089414639761565182862536 9

     7920827223758251185210916864000000000000000000000000 00

(c3) 101! / % ;

(d3)                                    101
```

MACSYMA has predicates and functions whose meaning is obvious but the application of which can be very long. That is true of the predicate of primality **primep** or of the factorisation function **factor**.

```
(c4) primep(12);

(d4)                                   false
```

```
(c5)  primep(216091);
```

$$(d5) \qquad\qquad\qquad\qquad\qquad\qquad true$$

```
(c6)  factor(2**63-1);
```

$$(d6) \qquad\qquad\qquad\qquad 7^2 \ \ 73 \ \ 127 \ \ 337 \ \ 92737 \ \ 649657$$

```
(c7)  large_prime : 2**216091 - 1 $
```

Here is an example of a loop with the instruction **for**. When the loop is finished, MACSYMA prints "done". That is, say, the (not very interesting) value corresponding to the instruction.

```
(c8)  for a:1 thru 5 do
    for b:1   thru 5 do
    if gcd(a,b)=1
    then print("a=",a,", b=", b ,", ",factor(b**6 + 3*a**6));
```

```
a= 1 , b= 1 ,    2²
a= 1 , b= 2 ,    67
a= 1 , b= 3 ,    2²  3 61
a= 1 , b= 4 ,    4099
a= 1 , b= 5 ,    2²  3907
a= 2 , b= 1 ,    193
a= 2 , b= 3 ,    3 307
a= 2 , b= 5 ,    15817
a= 3 , b= 1 ,    2²  547
a= 3 , b= 2 ,    2251
a= 3 , b= 4 ,    61 103
a= 3 , b= 5 ,    2²  61 73
a= 4 , b= 1 ,    12289
a= 4 , b= 3 ,    3 4339
a= 4 , b= 5 ,    103 271
a= 5 , b= 1 ,    2²  11719
a= 5 , b= 2 ,    73 643
a= 5 , b= 3 ,    2²  3 3967
a= 5 , b= 4 ,    50971
```

$$(d8) \qquad\qquad\qquad\qquad\qquad done$$

```
(c9)  for p:2 thru 3600 do
  if primep(p) and (remainder(2**(p-1), p**2) =1)
  then display(p);
```

$$p = 1093$$

$$p = 3511$$

(d9)   done

```
(c10) for n:2 thru 100 do
  for p:1 thru (n-1) do
  (q: n-p,
   if (gcd(p,q)=1 and integerp((4*p**3+27*q**2)**(1/2)))
   then display(p,q)
  );
```

$$p = 1$$

$$q = 10$$

$$p = 13$$

$$q = 34$$

$$p = 33$$

$$q = 32$$

$$p = 69$$

$$q = 22$$

(d10)   done

Examples of calculations with rational numbers:

(c11) s:0 $

(c12) for i:1 thru 100 do s:s+1/i;

(d12)   done

(c13) s;

$$(d13) \quad \frac{14466636279520351160221518043104131447711}{2788815009188499086581352357412492142272}$$

(c14) s: sum(1/i, i, 1, 100);

$$(d14) \quad \frac{14466636279520351160221518043104131447711}{2788815009188499086581352357412492142272}$$

Obviously MACSYMA can also calculate with real numbers with whatever precision the machine offers:

```
(c15)  216091*0.69314718;
```

```
(d15)                            149782.87
```

```
(c16)  a:123.897 * 1234.6 /(1.0 -345.765e-03);
```

```
(d16)                            233804.73
```

But thanks to the function **bfloat**, one can go beyond the precision and the range of the numbers offered by the hardware. In MACSYMA, such large real numbers are written syntactically: *n.nnnnnnnb* $\pm$ *nn*. The default precision is the value of the symbol **fpprec**.

```
(c17)  fpprec;
```

```
(d17)                               16
```

```
(c18)  bfloat(%pi);
```

```
(d18)                     3.141592653589793b0
```

```
(c19)  big:1.123424213423452515135b53*2.23423458725412745
           124523423544b64
```

```
(d19)                    2.50999323378944b117
```

```
(c20)  fpprec:36;
```

```
(d20)                               36
```

```
(c21)  bfloat(%pi);
```

```
(d21)             3.14159265358979323846264338327950288b0
```

```
(c22)  big:1.123424213423452515135b53*2.234234587254127451 2452
           3423544b64
```

```
(d22)          2.50999323378944021829124202358948023b117
```

```
(c23)  bfloat(s);
```

```
(d23)          5.1873775176396202608051176756582531 6b0
```

The following example shows the slow convergence of the series $\sum \frac{1}{n^2}$ to its sum: $\frac{\pi^2}{6}$.

```
(c24) s2: sum(1/i**2, i, 1, 200)$

(c25) float(%);

(d25)                              1.6399465

(c26) float(%pi**2/6);

(d26)                              1.6449341
```

We can use this classic problem to demonstrate function definition in MACSYMA. We shall calculate π by Méchain's formula:

$$\frac{\pi}{4} = 4 \arctan \frac{1}{5} - \arctan \frac{1}{239} \quad :$$

the expansion of which is:

$$\pi = 16 \sum_{p=0}^{\infty} \frac{(-1)^p}{(2p+1).5^{2p+1}} - 4 \sum_{p=0}^{\infty} \frac{(-1)^p}{(2p+1).239^{2p+1}}.$$

The input parameter n of the function is the number of terms chosen in the first series. We know that the error in such a series is less than the first term omitted and has the same sign. The calculation of the second series is done in such a way that the error is less than the error in the first series.

```
(c27) approx_pi (n) := block([s1,s2,isgn,isgn1,err1,
                              err2,err,i,j,sav],
    s1: 0, isgn: 1, sav: fpprec, fpprec: 3,
    for i:1   thru n do
    ( j: 2*i -1,
      s1: s1 + isgn /(j * 5**j), isgn: -isgn),
    err1: 16/((2*n+1)*5**(2*n+1)), isgn1: isgn,
    s1: 16*s1,
    s2: 0, isgn: 1,
    for i:1     step 1 while true do
    ( j: 2*i-1,
      err2: 1/(j * 239**j ),
      if (4*err2 - err1)<=0 then return(done),
      s2: s2 + isgn * err2,
      isgn: -isgn ),
    err:bfloat(isgn1*(err1 + 4* err2)), display(err),
    fpprec:sav,
    return (s1 - 4*s2) )$
```

We see that there is great similarity with the ALGOL-like programming languages. On the left of the symbol := is the name of the function followed by the list of the formal parameters. On the right we put the definition of the function in the form of a simple expression, or by using the symbol **block** in more complicated cases such as this. The instruction **return** has different meanings depending on the context: in the loop **for**, it makes us leave the loop, elsewhere it makes us leave the function, and the value of its argument is the value returned.

```
(c28) approx_pi(12);
```

$$err = 2.15b-18$$

```
        11164832817698171312722467003366646997915 2
(d28)   ---------------------------------------------
        35538766634625558164751756191253662109375
```

```
(c29) bfloat(%);
```

```
(d29)        3.141592653589793236392015619430860760
```

```
(c30) bfloat(%pi)- % ;
```

```
(d30)        2.0706277638486421844281762600212677460-18
```

```
(c31) sqrt(3.0);
```

```
(d31)                       1.7320508
```

```
(c32) fpprec:45$
```

```
(c33) bfloat(sqrt(3));
```

```
(d33) 1.732050807568877293527446341505872366942805250
```

```
(c34) log10_of_2: bfloat(log(2))/bfloat(log(10))$
```

```
(c35) integer(216091 * log10_of_2);
```

```
(d35)                         65049
```

```
(c36) bfloat(large_prime);
```

```
(d36) 7.460931030646613436873395794005114895402287540 b65049
```

We end with some examples with complex numbers. We shall explain later why we use the function **expand**.

```
(c37) z1: 4 + %i*19;
```

```
(d37)                     19 %i + 4
```

```
(c38) z2: 3 + 17 * %i;
```

```
(d38)                     17 %i + 3
```

(c39) z1+z2;

(d39) $36 \%i + 7$

(c40) z1*z2;

(d40) $(17 \%i + 3) (19 \%i + 4)$

(c41) expand(%);

(d41) $125 \%i - 311$

(c42) z1/z2;

(d42) $\dfrac{19 \%i + 4}{17 \%i + 3}$

(c43) rectform(%);

(d43) $\dfrac{335}{298} + \dfrac{11 \%i}{298}$

1.5.2 Polynomials and rational fractions

There is no need to expound the importance of polynomials and rational fractions here as they occur naturally in algebraic calculations. MACSYMA permits a certain number of manipulations on polynomials and rational fractions. The names of the corresponding commands are easily understood.

(c1) p0:2*x**3 + 7*x +9;

(d1) $2 x^3 + 7 x + 9$

(c2) p1: (a*x**2-y**3*z)*(y+b*z**2)*(x-y-2*z);

(d2) $(- 2 z - y + x) (a x^2 - y^3 z) (b z^2 + y)$

MACSYMA does not expand the products of sums systematically. But there are available functions such as **expand**, `ratexpand` or `rat`. The latter allows the user, *inter alia*, to arrange the polynomials in the desired order. The sign "/R/" in front of the result indicates a special form of machine representation called *recursive*. Explanations are to be found in a later chapter (see the section "Polynomials in several variables").

(c3) p1:expand(p1);

$$2 b y^3 z^4 + b y^4 z^3 - b x y^3 z^3 - 2 a b x^2 z^3 + 2 y^4 z^2$$
$$- a b x^2 y z^2 + a b x^3 z^2 + y^5 z - x y^4 z - 2 a x^2 y z$$
$$- a x^2 y^2 + a x^3 y$$

(d3)

(c4) p11:rat(p1,x,y,z);

(d4)/R/
$$2 b y^3 z^4 + (b y^4 - b x y^3 - 2 a b x^2) z^3$$
$$+ (2 y^4 - a b x^2 y + a b x^3) z^2 + (y^5 - x y^4 - 2 a x^2 y) z$$
$$- a x^2 y^2 + a x^3 y$$

(c5) p12:rat(p1,b,a,z,y,x);

(d5)/R/
$$(a y^2 + b a z^3) x + (- a y^2 + (- b a z^2 - 2 a z) y$$
$$- 2 b a z) x^3 + (- z y^4 - b z^3 y^3) x + z y^5$$
$$+ (b z^2 + 2 z^4) y^4 + 2 b z^4 y^3$$

(c6) p2:factor(p11);

(d6) $$(2 z + y - x) (y^3 z - a x^2) (b z^2 + y)$$

(c7) y0: 7*x**3-3*x+a;

(d7) $$7 x^3 - 3 x + a$$

(c8) y1: x**2-(a-1)*x +b;

(d8) $$x^2 - (a - 1) x + b$$

(c9) q1:y0/y1;

(d9) $$\dfrac{7 x^3 - 3 x + a}{x^2 - (a - 1) x + b}$$

(c10) quotient(y0,y1);

(d10) $7 x + 7 a - 7$

(c11) remainder(y0,y1);

(d11) $(- 7 b + 7 a^2 - 14 a + 4) x + (7 - 7 a) b + a$

(c12) remainder(x**4-a*x**3+(a+1)*x**2-2*x+3*a-2, x**2-3*x+2);

(d12) $(16 - 4 a) x + 7 a - 18$

(c13) num(q1);

(d13) $7 x^3 - 3 x + a$

(c14) denom(q1);

(d14) $x^2 - (a - 1) x + b$

(c15) y2: expand((x-a)**2*(y-b)*(z-c+1)**2);

(d15) $x^2 y^2 z - 2 a x y^2 z + a^2 y^2 z - b x^2 z + 2 a b x z$

$- a^2 b z - 2 c x^2 y^2 z + 2 x^2 y^2 z + 4 a c x y z - 4 a x y z$

$- 2 a^2 c y z + 2 a^2 y z + 2 b c x^2 z - 2 b x^2 z$

$- 4 a b c x z + 4 a b x z + 2 a^2 b c z - 2 a^2 b z + c^2 x^2 y^2$

$- 2 c x^2 y + x^2 y - 2 a c^2 x y + 4 a c x y - 2 a x y$

$+ a^2 c^2 y - 2 a^2 c y + a^2 y - b c^2 x^2 + 2 b c x^2 - b x^2$

$+ 2 a b c^2 x - 4 a b c x + 2 a b x - a^2 b c^2 + 2 a^2 b c$

$- a^2 b$

(c16) y3: expand((x-a)*(z-c+1));

(d16) $x z - a z - c x + x + a c - a$

(c17) gcd(y2,y3);

(d17) $(x - a) z + (1 - c) x + a c - a$

We see that the function **gcd** has indeed found the g.c.d. of the expanded polynomials without trying to factorise the result.

(c18) `factor(%);`

(d18) $(x - a) (z - c + 1)$

Similarly a rational fraction is not systematically simplified by the g.c.d. of its terms. To get this simplification we need to call a command such as **ratsimp**.

(c19) `q3:y2/y3;`

$$(d19) \quad (x^2 y z^2 - 2 a x y z^2 + a^2 y z^2 - b x^2 z^2 + 2 a b x z^2$$
$$- a^2 b z^2 - 2 c x^2 y z + 2 x^2 y z + 4 a c x y z - 4 a x y z$$
$$- 2 a^2 c y z + 2 a^2 y z + 2 b c x^2 z - 2 b x^2 z$$
$$- 4 a b c x z + 4 a b x z + 2 a^2 b c z - 2 a^2 b z + c^2 x^2 y^2$$
$$- 2 c x^2 y + x^2 y - 2 a c^2 x y + 4 a c x y - 2 a x y$$
$$+ a^2 c^2 y - 2 a^2 c y + a^2 y - b c^2 x^2 + 2 b c x^2 - b x^2$$
$$+ 2 a b c^2 x - 4 a b c x + 2 a b x - a^2 b c^2 + 2 a^2 b c$$
$$- a^2 b)/(x z - a z - c x + x + a c - a)$$

(c20) `ratsimp(y2/y3);`

(d20) $((x - a) y - b x + a b) z + ((1 - c) x + a c - a) y$
$$+ (b c - b) x - a b c + a b$$

The treatment of polynomials must have a way of getting the maximum and minimum power of a variable, as well as the coefficient of a term such as x^i:

(c21) `hipow(y2,c);`

(d21)   2

(c22) coeff(y2,z,1);

$$(d22) \quad - 2\,c\,x^2\,y + 2\,x^2\,y + 4\,a\,c\,x\,y - 4\,a\,x\,y - 2\,a^2\,c\,y$$

$$+ 2\,a^2\,y + 2\,b\,c\,x^2 - 2\,b\,x^2 - 4\,a\,b\,c\,x + 4\,a\,b\,x$$

$$+ 2\,a^2\,b\,c - 2\,a^2\,b$$

We can also go on to the elimination of variables between two equations: $y0 = 0$, $y1 = 0$ by the method of the *resultant*. (See the Appendix.)

(c23) resultant(y0,y1,x);

$$(d23) \quad 49\,b^3 + 42\,b^2 + a\,(63\,b - 4) - 12\,b + a^2\,(19 - 42\,b)$$

$$+ 7\,a^4 - 21\,a^3$$

To exploit these different possibilities, we write a function which lets us express a symmetric function (limited here to *three variables*) in terms of the fundamental symmetric functions, called $p1$, $p2$, $p3$. We use Waring's method [Dubreuil, l963].

```
(c24) waring3(vs):=
   block([ca,cb,cc,ha,hb,hc,vp1,vp2,vp3,sn,vpn],
   vp1:a+b+c,
   vp2:a*b+b*c+c*a,
   vp3:a*b*c,
   vpn: vs, sn:0,
   while(not(vpn=0)) do
     ( ca: cb : cc : ha : hb : hc : 0,
       ha : hipow (vpn,a),
       ca : coeff( vpn, a**ha),
       if (not (integerp (ca))) then
           (hb:hipow (ca,b),
            cb:coeff(ca,b**hb),
            if (not (integerp (cb))) then
                (hc:hipow (cb,c),
                 cc:coeff (cb,c**hc))
            else cc:cb )
         else cc:ca ,
       ha:ha-hb, hb:hb-hc,
       vpn: expand(vpn -  cc * vp1**ha * vp2**hb * vp3**hc),
       sn:sn + cc * p1**ha * p2**hb * p3**hc
       ),
       return(rat(sn,p3,p2,p1)) ) $
```

The command **kill** lets the variable be reset to the "atomic" state. In fact, at that moment it is essential that no value is assigned to the symbols $p1$, $p2$, $p3$. These symbols are not assigned any value during execution. Therefore we only need to use the command **kill** once.

(c25) kill(p1,p2,p3);

(d25)   done

(c26) waring3(a**3+b**3+c**3);

$$
\text{(d26)/R/} \qquad\qquad p1^3 \;-\; 3\,p2\,p1 \;+\; 3\,p3
$$

(c27) s17: waring3(a**17+b**17+c**17);

$$
\text{(d27)/R/}\; p1^{17} \;-\; 17\,p2\,p1^{15} \;+\; 17\,p3\,p1^{14} \;+\; 119\,p2^2\,p1^{13}
$$

$$
-\; 221\,p3\,p2\,p1^{12} \;+\; (-\,442\,p2^3 \;+\; 102\,p3^2\,)\,p1^{11}
$$

$$
+\; 1122\,p3\,p2^2\,p1^{10} \;+\; (935\,p2^4 \;-\; 935\,p3^2\,p2)\,p1^9
$$

$$
+\; (-\,2805\,p3\,p2^3 \;+\; 255\,p3^3\,)\,p1^8
$$

$$
+\; (-\,1122\,p2^5 \;+\; 3060\,p3^2\,p2^2\,)\,p1^7
$$

$$
+\; (3570\,p3\,p2^4 \;-\; 1428\,p3^3\,p2)\,p1^6
$$

$$
+\; (714\,p2^6 \;-\; 4284\,p3^2\,p2^3 \;+\; 238\,p3^4\,)\,p1^5
$$

$$
+\; (-\,2142\,p3\,p2^5 \;+\; 2380\,p3^3\,p2^2\,)\,p1^4
$$

$$
+\; (-\,204\,p2^7 \;+\; 2380\,p3^2\,p2^4 \;-\; 595\,p3^4\,p2)\,p1^3
$$

$$
+\; (476\,p3\,p2^6 \;-\; 1190\,p3^3\,p2^3 \;+\; 51\,p3^5\,)\,p1^2
$$

$$
+\; (17\,p2^8 \;-\; 357\,p3^2\,p2^5 \;+\; 255\,p3^4\,p2^2\,)\,p1 \;-\; 17\,p3\,p2^7
$$

$$
+\; 85\,p3^3\,p2^4 \;-\; 17\,p3^5\,p2
$$

(c28) discri3: waring3(expand((a-b)**2 * (a-c)**2 * (b-c)**2));

$$(d28)/R/ \; - \; 4 \; p3 \; p1^{3} \; + \; p2^{2} \; p1^{2} \; + \; 18 \; p3 \; p2 \; p1 \; - \; 4 \; p2^{3} \; - \; 27 \; p3^{2}$$

The preceding method may be general; it is not however the most efficient, especially for expressing sums of like powers. In the case we are studying — for three variables only — as soon as we have obtained

$$s_0 = 3, s_1 = p_1, s_2 = p_1^2 - 2p_2$$

we can calculate s_n by the following recurrence:

$$s_{n+1} = p_1 s_n - p_2 s_{n-1} + p_3 s_{n-2}.$$

That allows us to introduce indexed variables, which in this simple case can be used in MACSYMA without prior declaration. The calculation of s_{17} on line $(c27)$ takes several minutes on the Honeywell computer DPS8-70. That of $det[17]$ takes only a few seconds by the present method.

```
(c29) det[0]: 3 $
```

```
(c30) det[1]: p1 $
```

```
(c31) det[2]: p1**2 - 2*p2 $
```

```
(c32) for n:3 thru 17 do det[n]:expand(p1*det[n-1]-p2*det[n-2]
                                    + p3*det[n-3]);
```

(d32)   done

```
(c33) det[4];
```

(d33) $- 2 \; p1 \; p3 + 2 \; p2^{2} - 4 \; p1^{2} \; p2 + p1^{4}$

```
(c34) expand(s17 -det[17]);
```

0

Using this type of indexed variables is especially efficient in the following recursive mechanism of definition in an array. The calculation of an element will only be done once and the result will be stored.

```
(c35) deter[n] := if n=0 then 3 else
           if n=1 then p1 else
           if n=2 then p1**2 - 2*p2 else
           rat(p1*deter[n-1]-p2*deter[n-2]+p3*deter[n-3],p3,p2,p1)$
```

```
(c36) deter[3];
```

$$p1^3 - 3\ p2\ p1 + 3\ p3$$

(d36)/R/

```
(c37) deter[4];
```

$$p1^4 - 4\ p2\ p1^2 + 4\ p3\ p1 + 2\ p2^2$$

(d37)/R/

```
(c38) sav7 : deter[7] $

(c39) deter[5];
```

$$p1^5 - 5\ p2\ p1^3 + 5\ p3\ p1^2 + 5\ p2^2\ p1 - 5\ p3\ p2$$

(d39)/R/

We end with an example of factorisation of a polynomial in several variables. The following example takes 4 minutes on the Honeywell computer DPS8-70. Although factorisation algorithms are not discussed until Chapter 4, the reader can already see that factorisation is an expensive process.

```
(c40) for n:2 thru 10 do print(factor((u^n+v^n)^n - (u^n-v^n)^n));
```

$$4\ u^2\ v^2$$

$$2\ v^3\ (v^6 + 3\ u^6)$$

$$8\ u^4\ v^4\ (v^8 + u^8)$$

$$2\ v^5\ (v^{20} + 10\ u^{10}\ v^{10} + 5\ u^{20})$$

$$4\ u^6\ v^6\ (v^{12} + 3\ u^{12})\ (3\ v^{12} + u^{12})$$

$$2\ v^7\ (v^{42} + 21\ u^{14}\ v^{28} + 35\ u^{28}\ v^{14} + 7\ u^{42})$$

$$16\ u^8\ v^8\ (v^{16} + u^{16})\ (v^{32} + 6\ u^{16}\ v^{16} + u^{32})$$

$$2\ v^9\ (v^{18} + 3\ u^{18})\ (v^{54} + 33\ u^{18}\ v^{36} + 27\ u^{36}\ v^{18} + 3\ u^{54})$$

$$4\ u^{10}\ v^{10}\ (v^{40} + 10\ u^{20}\ v^{20} + 5\ u^{40})\ (5\ v^{40} + 10\ u^{20}\ v^{20} + u^{40})$$

(d40)   done

1.5.3 Matrix calculation

Vectors and matrices, constantly used in numerical calculations, are obviously necessary in Computer Algebra. The user has several means available for defining these objects.

(1) Explicit declaration:

```
(c1) m0: matrix([1, 2, 3], [4, 5, 6], [7, 8, 9]);
```

```
              [ 1   2   3 ]
              [           ]
(d1)          [ 4   5   6 ]
              [           ]
              [ 7   8   9 ]
```

```
(c2) v1: [a, b, c];
```

```
(d2)              [ a   b   c ]
```

```
(c3) v2:transpose(v1);
```

```
                  [ a ]
                  [   ]
(d3)              [ b ]
                  [   ]
                  [ c ]
```

(2) Declaration with interactive assignment:

```
(c4) m1:entermatrix(3,3);
```

```
Is the matrix   1. Diagonal   2. Symmetric   3. Antisymmetric
                4. General
Answer 1, 2, 3 or 4
```

```
2;
```

```
Row 1 Column 1:  u;
```

```
Row 1 Column 2:  v;
```

```
Row 1 Column 3:  w;
```

```
Row 2 Column 2:  x;
```

```
Row 2 Column 3:  y;
```

```
Row 3 Column 3:  z;
```

```
Matrix entered.
```

```
            [ u   v   w ]
            [           ]
(d4)        [ v   x   y ]
            [           ]
            [ w   y   z ]
```

(3) Definition via a generating function for the elements:

(c5) h[i,j]:=1/(i+j-1)$

(c6) hilbert4: genmatrix(h,4,4);

```
        [     1   1   1 ]
        [ 1   -   -   - ]
        [     2   3   4 ]
        [             ]
        [ 1   1   1   1 ]
        [ -   -   -   - ]
        [ 2   3   4   5 ]
(d6)    [             ]
        [ 1   1   1   1 ]
        [ -   -   -   - ]
        [ 3   4   5   6 ]
        [             ]
        [ 1   1   1   1 ]
        [ -   -   -   - ]
        [ 4   5   6   7 ]
```

The elements are selected by a simple method:

(c7) s0: m0[1,2];

(d7)   2

(c8) s0: m1[1,1] + v1[2] * hilbert4 [3,3] ;

```
                        b
(d8)            u + -
                        5
```

(c9) v3: row(hilbert4, 3);

```
            [ 1   1   1   1 ]
(d9)        [ -   -   -   - ]
            [ 3   4   5   6 ]
```

(c10) v4: col (hilbert4, 2);

```
                        [ 1 ]
                        [ - ]
                        [ 2 ]
                        [   ]
                        [ 1 ]
                        [ - ]
                        [ 3 ]
(d10)                   [   ]
                        [ 1 ]
                        [ - ]
                        [ 4 ]
                        [   ]
                        [ 1 ]
                        [ - ]
                        [ 5 ]
```

One can also modify one element of the matrix and this modifies the matrix itself.

(c11) m0[2,2] : x*y;

(d11)   x y

(c12) m0;

```
                        [ 1    2    3 ]
                        [             ]
(d12)                   [ 4   x y   6 ]
                        [             ]
                        [ 7    8    9 ]
```

(c13) m0[2,2]: 5$

The operator of matrix multiplication is the dot ".". The usual arithmetical operators work element by element.

(c14) p: v3 . v4;

```
                          1
(d14)                     -
                          3
```

(c15) v6: m0 . v2;

```
                        [  3 c + 2 b + a  ]
                        [                 ]
(d15)                   [ 6 c + 5 b + 4 a ]
                        [                 ]
                        [ 9 c + 8 b + 7 a ]
```

(c16) m2: 2 * m0;

```
         [ 2    4    6  ]
         [              ]
(d16)    [ 8    10   12 ]
         [              ]
         [ 14   16   18 ]
```

(c17) m22: m0 * m0;

```
         [ 1    4    9  ]
         [              ]
(d17)    [ 16   25   36 ]
         [              ]
         [ 49   64   81 ]
```

(c18) msquare: m0 . m0;

```
         [ 30   36    42   ]
         [                 ]
(d18)    [ 66   81    96   ]
         [                 ]
         [ 102  126   150  ]
```

(c19) mp1: 1+ m0;

```
         [ 2    3    4  ]
         [              ]
(d19)    [ 5    6    7  ]
         [              ]
         [ 8    9    10 ]
```

There are also an exponential operator `^^`, and functions for obtaining the *determinant* and the *characteristic polynomial*.

(c20) invhilb4: hilbert4^^(-1);

```
         [  16     - 120     240   - 140    ]
         [                                  ]
         [ - 120    1200    - 2700   1680   ]
(d20)    [                                  ]
         [  240   - 2700    6480   - 4200   ]
         [                                  ]
         [ - 140    1680    - 4200   2800   ]
```

(c21) polc: rat(charpoly(m1,lambda),lambda);

(d21)/R/ - lambda^3 + (z + x + u) lambda^2

+ ((- x - u) z + y^2 - u x + w^2 + v^2) lambda + (u x - v^2) z

$$- u\, y^2 + 2\, v\, w\, y - w^2\, x$$

We give several examples of MACSYMA's ability to deal with matrices obtained by permuting circularly the elements of a row vector

$$(a, b, \ldots).$$

We shall *check* one property of these matrices. Consider the expansion of the binomial $(u + v)^n$. If we write

$$a = u^n + v^n, b = C_n^1 u.v^{n-1}, c = C_n^2 u^2.v^{n-2}, \ldots,$$

the determinant of the matrix is:

$$(u^n - (-1)^n.v^n)^n.$$

This is related to the fact that the determinant of the matrix resulting from permuting cyclically $(u, v, 0, 0, \ldots)$ is

$$(u^n - (-1)^n.v^n)$$

(c22) p3: matrix([a, b, c], [c, a, b], [b, c, a]);

$$(d22) \qquad \begin{bmatrix} a & b & c \\ c & a & b \\ b & c & a \end{bmatrix}$$

(c23) determinant(p3);

$$(d23) \qquad c\,(c^2 - a\,b) + a\,(a^2 - b\,c) - b\,(a\,c - b^2)$$

(c24) ratsimp(%);

$$(d24) \qquad a^3 + b^3 + c^3 - 3\,a\,b\,c$$

Such a determinant is obviously divisible by $a + b + c$:

(c25) factor(d24);

$$(d25) \qquad (c + b + a)\,(c^2 - b\,c - a\,c + b^2 - a\,b + a^2)$$

If in the cases 5, 7, 11, we put a limit of 3 parameters in the initial vector, we find the following determinants:

$$a^5 + b^5 + c^5 - 5abc(b^2 - ac)$$

$$a^7 + b^7 + c^7 - 7abc(b^2 - ac)^2$$

$$a^{11} + b^{11} + c^{11} - 11abc(b^2 - ac)(b^6 - 3ab^4c + 4a^2b^2c^2 - a^3c^3)$$

Let us go back to the initial matrix $(3, 3)$:

(c26) a:u**3+v**3$

(c27) b: 3*u**2*v$

(c28) c: 3*u*v**2$

(c29) p3: matrix([a,b,c],[c,a,b],[b,c,a]);

```
       [  3    3     2           2   ]
       [ v  + u   3 u  v     3 u v   ]
       [                            ]
(d29)  [     2      3    3      2    ]
       [ 3 u  v    v  + u    3 u  v  ]
       [                            ]
       [    2           2    3    3  ]
       [ 3 u  v    3 u v    v  + u   ]
```

(c30) factor(determinant(p3));

```
                      3    2           2  3
(d30)          (v + u)  (v  - u v + u  )
```

What follows is simply a device to get the result in the expected form:

(c31) expand(% ** (1/3)) **3 ;

```
               3    3 3
(d31)         (v  + u )
```

Let us study cases 4 and 5 in turn:

(c32) a: u**4+v**4$

(c33) b: 4*u**3*v$

(c34) c:6*u**2*v**2$

(c35) d:4*u*v**3$

```
(c36) p4: matrix([a,b,c,d],
                 [d,a,b,c],
                 [c,d,a,b],
                 [b,c,d,a]) $
```

Instead of using the contrivance of line $c31$, it is better to use the function **sqfr** which is analogous to **factor** but which brings out the factors common to the polynomial and its derivatives.

```
(c37) sqfr(determinant(p4));
```

$$(d37) \qquad (v^4 - u^4)^4$$

```
(c38) a:u**5 + v**5$

(c39) b:5*u**4*v$

(c40) c:10*u**3*v**2$

(c41) d:10*u**2*v**3$

(c42) e:5*u*v**4$

(c43) p5: matrix([a,b,c,d,e],
                 [e,a,b,c,d],
                 [d,e,a,b,c],
                 [c,d,e,a,b],
                 [b,c,d,e,a]) $

(c44) sqfr(determinant(p5));
```

$$(d44) \qquad (v^5 + u^5)^5$$

We end by writing a function which gives the inverse of a matrix by the Souriau method. We express the result in the form of the adjoint of the initial matrix. The determinant is the value of the **global** variable *det*. We shall see that in $c48$ and $c49$ the value of *det* is different. The user must check carefully such so-called *side-effects*.

```
(c45) souriau (a):=block([a1,b,i,j,q,n],
   n:length(a),
   a1:copymatrix(a),
   for j:1 thru n do (
       q:0,for i:1 thru n do q:q+a1[i,i],q:ratexpand(q/j),
       if j<n then (
           for i:1 thru n do a1[i,i]:ratexpand(a1[i,i]-q),
           if j=(n-1) then b:copymatrix(a1),
           a1: ratexpand( a.a1)
           )
               ),
   det: (-1)^(n+1)*q,
   return(b) ) $
```

(c46) adjm1: souriau(m1);

$$
\text{(d46)} \quad
\begin{bmatrix}
x z - y^2 & w y - v z & v y - w x \\
w y - v z & u z - w^2 & v w - u y \\
v y - w x & v w - u y & u x - v^2
\end{bmatrix}
$$

(c47) det;

$$
\text{(d47)} \qquad u x z - v^2 z - u y^2 + 2 v w y - w^2 x
$$

(c48) expand(det - determinant(m1));

$$
\text{(d48)} \qquad\qquad 0
$$

(c49) invhilb4 - souriau(hilbert4) / det;

$$
\text{(d49)} \quad
\begin{bmatrix}
0 & 0 & 0 & 0 \\
0 & 0 & 0 & 0 \\
0 & 0 & 0 & 0 \\
0 & 0 & 0 & 0
\end{bmatrix}
$$

To solve a linear system, we can use the command *solve* in the form

$$
solve([eq_1, eq_2, \ldots], [x_1, x_2, \ldots])
$$

where the eq_i are the equations (here the right hand side, assumed equal to 0, can be omitted) and where the x_i are the unknowns. There follow some examples of unstable numerical systems. The result of the function **solve** is given in vectorial form. We shall see later how to extract the interesting values. The reader should note how much precise calculation over **Q** can be of assistance in cases of instability.

(c1) eq11: 3*x+4*y-7;

(d1)   4 y + 3 x - 7

(c2) eq12: 3*x+400001/100000 *y-700001/100000;

(d2)
$$\frac{400001\ y}{100000} + 3\ x - \frac{700001}{100000}$$

(c3) solve([eq11,eq12], [x,y]);

(d3)   [[x = 1, y = 1]]

(c4) eq13: 3*x+ 399999/100000 *y -700004/100000;

(d4)
$$\frac{399999\ y}{100000} + 3\ x - \frac{175001}{25000}$$

(c5) solve([eq11,eq13], [x,y]);

(d5)
$$[[x = \frac{23}{3}, y = - 4]]$$

(c6) eq14: 3*x+3999992/1000000 *y - 7000042/1000000;

(d6)
$$\frac{499999\ y}{125000} + 3\ x - \frac{3500021}{500000}$$

(c7) solve([eq11,eq14],[x,y]);

(d7)
$$[[x = \frac{28}{3}, y = - \frac{21}{4}]]$$

Of course one can also solve systems where the matrix is symbolic. Here the equations are written explicitly.

(c8) e1: a*(r+s) +3*b*(r-s) = u+v;

(d8)   a (s + r) + 3 b (r - s) = u + v

(c9) e2: a*(r-s) -b*(r+s) = u-v;

(d9)   - b (s + r) + a (r - s) = u - v

(c10) e3: 1-3*w = u + v;

(d10) $- 3 \ w \ + 1 = u + v$

(c11) e4:a**2+3*b**2-3*z = r+s;

(d11) $- 3 \ z + 3 \ b^2 \ + a^2 = r + s$

(c12) solve([e1,e2,e3,e4],[u,v,r,s]);

(d12) [[u = $- (- 3 \ a^2 \ z + b^2 \ (6 \ a^2 \ - 9 \ z) + b \ (9 \ w - 3)$

$+ a \ (3 \ w - 1) + 9 \ b^4 \ + a^4 \ )/(6 \ b),$

$v = (- 3 \ a^2 \ z + b^2 \ (6 \ a^2 \ - 9 \ z) + a \ (3 \ w - 1) + b \ (3 - 9 \ w)$

$+ 9 \ b^4 \ + a^4 \ )/(6 \ b),$

$r = (3 \ a^2 \ z + b \ (3 \ a^3 \ - 9 \ z) - 3 \ w + 9 \ b^2 \ - 3 \ a \ b - a^3 \ + 1)$

$/(6 \ b),$

$s = (- 3 \ a^2 \ z + b \ (3 \ a^3 \ - 9 \ z) + 3 \ w + 9 \ b^2 \ + 3 \ a \ b + a^3 \ - 1)$

$/(6 \ b)$

]]

1.5.4 Differentiation – Taylor series

Every Computer Algebra system contains a set, more or less complete, of commands connected with the concept of **differentiation**. These operations are closely linked to the **simplification** commands. Without good simplification tools, there is a risk of very quickly getting incomprehensible formulae.

```
(c1)  f1:2*(log(x**2-x+1)/6 + atan((2*x-1)/sqrt(3))/sqrt(3)

            -log(x+1)/3);
```

$$(d1) \quad 2\left(\dfrac{\log(x^2 - x + 1)}{6} + \dfrac{\operatorname{atan}\left(\dfrac{2x - 1}{\operatorname{sqrt}(3)}\right)}{\operatorname{sqrt}(3)} - \dfrac{\log(x + 1)}{3}\right)$$

```
(c2)  df1: diff(f1,x);
```

$$(d2) \quad 2\left(\dfrac{2}{3\left(\dfrac{(2x - 1)^2}{3} + 1\right)} + \dfrac{2x - 1}{6(x^2 - x + 1)} - \dfrac{1}{3(x + 1)}\right)$$

```
(c3)  df1:ratsimp(df1);
```

$$(d3) \qquad \dfrac{2x}{x^3 + 1}$$

MACSYMA has also the function

$$taylor(function, variable, initial, order)$$

The symbol "/T/" at the front of the result means that we are dealing with a truncated expansion. Operations on a result of this kind take into account the order attached to it. On line $c7$, we use the function subst(x, y, z) which substitutes x in place of y in z.

```
(c4)  q: (1- sqrt(1-e**2))/e;
```

$$(d4) \qquad \dfrac{1 - \operatorname{sqrt}(1 - e^2)}{e}$$

```
(c5)  taylor(q,e,0,10);
```

$$(d5)/T/ \qquad \dfrac{e}{2} + \dfrac{e^3}{8} + \dfrac{e^5}{16} + \dfrac{5e^7}{128} + \dfrac{7e^9}{256} + \ldots$$

```
(c6)  a3:(u-sin(u))/sin(u)**3;
```

(d6)
$$\frac{u - \sin(u)}{\sin^3(u)}$$

(c7) a3: subst(2*t/(1+t**2), sin(u),a3)$

(c8) a3:subst(2*atan(t),u,a3)$

(c9) a3t: taylor(a3,t,0,10);

$$(d9)/T/ \quad \frac{1}{6} + \frac{3\,t^2}{10} + \frac{4\,t^4}{35} - \frac{4\,t^6}{315} + \frac{4\,t^8}{1155} - \frac{4\,t^{10}}{3003} + \ldots$$

(c10) taylor(sqrt(1- (k*sin(x))**2),x,0,6);

$$(d10)/T/ \quad 1 - \frac{k^2\,x^2}{2} - \frac{(3\,k^4 - 4\,k^2)\,x^4}{24}$$

$$- \frac{(45\,k^6 - 60\,k^4 + 16\,k^2)\,x^6}{720} + \ldots$$

(c11) %*%;

$$(d11)/T/ \quad 1 - k^2\,x^2 + \frac{k^4\,x^4}{3} - \frac{(2\,k^2)\,x^6}{45} + \ldots$$

One can also work with the *differentiation operator.* In the first example we show how to obtain the successive derivatives of y with respect to x starting out from an implicit form $g(x, y)$. The command **depends** here means that g depends on x and y and that y depends on x. The function **part** lets us choose a part of the formula by a method linked to the tree representation of this formula in memory. We see that MACSYMA uses the same symbol **d** for the total derivative and for the partial derivative.

(c12) depends(g,[x,y],y,[x]);

(d12)   [g(x, y), y(x)]

(c13) diff(g,x);

$$(d13) \quad \frac{dg}{dy}\frac{dy}{dx} + \frac{dg}{dx}$$

(c14) solve(%,diff(y,x));

$$
\text{(d14)} \qquad [\frac{dy}{dx} = - \frac{\dfrac{dg}{dx}}{\dfrac{dg}{dy}}]
$$

(c15) dydx:part(%,1,2);

$$
\text{(d15)} \qquad - \frac{\dfrac{dg}{dx}}{\dfrac{dg}{dy}}
$$

(c16) diff(g,x,2);

$$
\text{(d16)} \quad \frac{dg}{dy}\frac{d^2 y}{dx^2} + \frac{dy}{dx}\left(\frac{d^2 g}{dy^2}\frac{dy}{dx} + \frac{d^2 g}{dx\,dy}\right) + \frac{d^2 g}{dx\,dy}\frac{dy}{dx} + \frac{d^2 g}{dx^2}
$$

(c17) subst(dydx,diff(y,x),%)$

(c18) solve(%,diff(y,x,2))$

(c19) d2ydx2:part(%,1,2);

$$
\text{(d19)} \qquad - \frac{\left(\dfrac{dg}{dx}\right)^2\dfrac{d^2 g}{dy^2} + \dfrac{d^2 g}{dx^2}\left(\dfrac{dg}{dy}\right)^2 - 2\dfrac{dg}{dx}\dfrac{d^2 g}{dx\,dy}\dfrac{dg}{dy}}{\left(\dfrac{dg}{dy}\right)^3}
$$

(c20) diff(g,x,3)$

(c21) subst(dydx,diff(y,x),subst(d2ydx2,diff(y,x,2),%))$

(c22) solve(%,diff(y,x,3))$

(c23) d3ydx3: part(%,1,2);

```
                  3                    2
          dg 3 dg d g        dg 3  d g 2
(d23)    ((--)  -- ---  - 3 (--)  (---)
          dx    dy   3        dx     2
                   dy                dy


              2                   2              2     3
        dg 2  d g  dg      dg d g  dg 2  d g     d g  dg 4
  + (9 (--)  ----- --  - 3 -- --- (--) ) ---  - --- (--)
        dx   dx dy dy      dx   2   dy    2      3   dy
                                 dx           dy    dx


            3            2   2
        dg  d g       d g  d g  dg 3
  + (3  -- ------  + 3 ----- ---) (--)
        dx   2         dx dy   2   dy
           dx dy               dx


             3                  2
        dg 2  d g       dg  d g  2   dg 2   dg 5
  + (- 3 (--)  ------  - 6 -- (-----) ) (--) )/(--)
         dx     2          dx  dx dy      dy     dy
              dx dy
```

(c24) remove([x,y],dependency)$

Once we have these formulae we may want to apply them to a concrete
example. First we replace g by the function $exp(x^2 + y^2)$, then we hand
over to the very general MACSYMA command **ev** the task of applying the
differentiation operators. Obviously we will recover the derivatives linked
to the function

$$x^2 + y^2 = constant.$$

(c25) fct0: subst(%e**(x**2+y**2)-1,g,dydx);

```
                    2    2
           d      y  + x
           -- (%e        - 1)
           dx
(d25)    - --------------------
                    2    2
           d      y  + x
           -- (%e        - 1)
           dy
```

(c26) ev(fct0,diff);

```
                    x
(d26)            - -
                    y
```

```
(c27)  fct1:subst(%e**(x**2+y**2)-1,g,d2ydx2)$
```

```
(c28)  ev(fct1,diff)$
```

```
(c29)  ratexpand(%);
```

$$
(d29) \qquad -\ \frac{1}{y}\ -\ -\ \frac{x^2}{\dfrac{3}{y}}
$$

```
(c30)  remove([g],dependency)$
```


The following example shows how to get the derivatives of a "function of a function" by using a notation for *composition*: $g \circ f$. We use artificially the *non-commutative multiplication* operator, i.e. the **dot**. The successive derivatives of the function g will be indicated by a superscript.

```
(c31)  depends(g,f,f,x);
```

$$
(d31) \qquad\qquad [g(f),\ f(x)]
$$

```
(c32)  d1:diff(g,x);
```

$$
(d32) \qquad\qquad \frac{df}{dx}\,\frac{dg}{df}
$$

```
(c33)  subst(g[1].f ,diff(g,f),d1);
```

$$
(d33) \qquad\qquad (g_1\ .\ f)\ \frac{df}{dx}
$$

```
(c34)  d2:diff(g,x,2);
```

$$
(d34) \qquad \left(\frac{df}{dx}\right)^2 \frac{d^2 g}{df^2} + \frac{d^2 f}{dx^2}\,\frac{dg}{df}
$$

```
(c35)  dd2:d2$
```

```
(c36)  for i:1 thru 2 do dd2: subst(g[i].f, diff(g,f,i), dd2)$
```

```
(c37)  dd2;
```

$$(d37) \qquad (g_1 \ . \ f) \ \frac{d^2 f}{dx^2} + (g_2 \ . \ f) \ (\frac{df}{dx})^2$$

(c38) d3:diff(g,x,3)\$

(c39) dd3:d3\$

(c40) for i:1 thru 3 do dd3:subst(g[i].f,diff(g,f,i),dd3)\$

(c41) dd3;

$$(d41) \quad (g_1 \ . \ f) \ \frac{d^3 f}{dx^3} + 3 \ (g_2 \ . \ f) \ \frac{df}{dx} \ \frac{d^2 f}{dx^2} + (g_3 \ . \ f) \ (\frac{df}{dx})^3$$

(c42) d5:diff(g,x,5)\$

(c43) dd5:d5\$

(c44) for i:1 thru 5 do dd5:subst(g[i].f,diff(g,f,i),dd5)\$

(c45) dd5;

$$(d45) \quad (g_1 \ . \ f) \ \frac{d^5 f}{dx^5} + 5 \ (g_2 \ . \ f) \ \frac{df}{dx} \ \frac{d^4 f}{dx^4} + 10 \ (g_2 \ . \ f) \ \frac{d^2 f}{dx^2} \ \frac{d^3 f}{dx^3}$$

$$+ 10 \ (g_3 \ . \ f) \ (\frac{df}{dx})^2 \ \frac{d^3 f}{dx^3} + 15 \ (g_3 \ . \ f) \ \frac{df}{dx} \ (\frac{d^2 f}{dx^2})^2$$

$$+ 10 \ (g_4 \ . \ f) \ (\frac{df}{dx})^3 \ \frac{d^2 f}{dx^2} + (g_5 \ . \ f) \ (\frac{df}{dx})^5$$

1.5.5 Simplification of formulae

The problem of simplification is a complex one, as we shall see later in the book. Modern systems. and especially MACSYMA, offer the user a set of switches and parameters which let him control the expansion of formulae, and thus prevent a brute force use of the commands from giving a very complicated result in which the information is completely "drowned". We saw earlier the use of the function **expand**. In addition to it there is a whole range of functions which enable us to carry out more delicate transformations: factorisation, limited expansions, use or not of the g.c.d., ordering of the variables, distribution of products, use or not of commutativity etc. The correct use of these switches and of the commands is a matter of practice. It is clear that brute force attack on a real problem can give rise to surprises: huge expansions, prohibitive time taken, exhaustion of memory .... It is hard to give exact rules, except *think before you act* and try out some simple examples to appreciate what is involved. The following few simple examples demonstrate some of the possibilities. The interested reader should consult the REDUCE or MACSYMA manual.

We have already seen the fu. .ion **subst** which performs a purely syntactic substitution. Thus in line $c2$, **subst** does not recognise the grouping $x+y$ in the third argument. On the other hand the function **ratsubst** analyses this third argument in more detail. Finally with the function **fullsub**, defined *in situ*, we see that the process can be used recursively.

```
(c2)  subst(a,x+y,x+z+y);
```

```
(d2)                        z + y + x
```

```
(c3)  ratsubst(a,x+y,y+z+x);
```

```
(d3)                        z + a
```

```
(c4)  ratsubst(b*a,a**2,a**4);
```

```
                       2  2
(d4)               a  b
```

```
(c5)  fullsub(u,v,w):= if w=(w2:ratsubst(u,v,w))
                       then w else fullsub(u,v,w2)$
```

```
(c6)  fullsub(b*a,a**2,a**4);
```

```
                         3
(d6)               a  b
```

With the system variables **maxposex** and **maxnegex**, we can control the expansion of expressions so as to restrain those which would run the risk of *exploding*.

```
(c7)  maxposex:4;
```

(d7)   4

(c8) p0:(a+b)**2 +(a+b)**3 +(a+b)**5;

$$(d8) \qquad (b + a)^5 + (b + a)^3 + (b + a)^2$$

(c9) p0:expand(p0);

$$(d9) \quad (b + a)^5 + b^3 + 3\,a\,b^2 + b^2 + 3\,a^2\,b + 2\,a\,b + a^3 + a^2$$

(c10) maxnegex:3;

(d10)   3

(c11) p2: 1/(x-a)**2 + 1/(x-b)**2 +a*b/(x-a-b)**4;

$$(d11) \qquad \frac{a\,b}{(x - b - a)^4} + \frac{1}{(x - b)^2} + \frac{1}{(x - a)^2}$$

(c12) p3:expand(p2);

$$(d12) \quad \frac{1}{x^2 - 2\,b\,x + b^2} + \frac{1}{x^2 - 2\,a\,x + a^2} + \frac{a\,b}{(x - b - a)^4}$$

The Boolean switch **exponentialize**, when set **true**, lets us put the trigonometrical and hyperbolic functions into exponential form.

(c13) exponentialize:true;

(d13)   true

(c14) z1:cosh(x);

$$(d14) \qquad \frac{\%e^x + \%e^{-x}}{2}$$

(c15) z0:tan(x);

$$(d15) \quad - \frac{\%i\,(\%e^{\%i\,x} - \%e^{-\%i\,x})}{\%e^{\%i\,x} + \%e^{-\%i\,x}}$$

The switch *exponentialize* or the function bearing the same name often prove useful in simplifying complicated trigonometrical expressions. Morley's theorem is an example*. Let ABC be a triangle. Let us take the trisectors of the three angles. The two trisectors of the angles B and C which are closest to the side BC intersect at A_1. In the same way we get B_1 and C_1. *The triangle $A_1B_1C_1$ is equilateral.* It is sufficient to prove that it is isosceles, that is: $A_1B_1 = A_1C_1$. Using the relations

$$AB = 2R\sin C, BC = 2R\sin A, CA = 2R\sin B$$

(R being the radius of the circumscribed circle) and the well-known formulae for resolution of triangles, one can express A_1B_1 and A_1C_1 as functions of R, B and C. The following commands take account of the homogeneity in R and of the obvious permutations. Here we see that the function **ev** allows substitution in *parallel*. The equality of the two lengths is shown in two steps. We first extract the numerators of the difference then apply the very general simplification function **radcan** to the exponential form of this numerator. By putting the value of the global parameter **ratdenomdivide** to **false**, we ensure that the resulting rational fraction will stay in the form of a fraction and not of a sum of fractions. This "brute force" calculation is long and difficult and the reader will undoubtedly prefer Coxeter's elegant proof [Coxeter, 1961], but it is still true that this is a characteristic example of the help that can be given by a Computer Algebra system.

```
(c3) ba1:   sin(a) * sin(c/3) / sin((b+c)/3);

                                c
                     sin(a) sin(-)
                                3
(d3)                 -------------
                         c + b
                     sin(-----)
                           3


(c4) bc1:  ev(ba1,a=c,c=a);

                       a
                 sin(-) sin(c)
                       3
(d4)             -------------
                     b + a
                 sin(-----)
                       3


(c5) ba1:  subst(%pi-b-c,a,ba1);
```

---

* D. Lazard suggested this example and the solution.

$$
\text{(d5)} \qquad \frac{\sin(\dfrac{c}{3})\ \sin(c + b)}{\sin(\dfrac{c + b}{3})}
$$

(c6) bc1: subst(%pi-b-c,a,bc1);

$$
\text{(d6)} \qquad \frac{\sin(\dfrac{-\ c\ -\ b\ +\ \%pi}{3})\ \sin(c)}{\sin(\dfrac{\%pi\ -\ c}{3})}
$$

(c7) a1c12:ba1^2+bc1^2-2*ba1*bc1*cos(b/3)$

(c8) a1b12:ev(a1c12,b=c,c=b)$

(c9) ratdenomdivide:false$

(c10) r:num(ratexpand(a1b12-a1c12))$

Time= 922 msec.

(c11) radcan(exponentialize(r));

Time= 68948 msec.

(d11)   0

An operator can also be applied to each element of a sum, thanks to the function **map**. In the following example we break down a fraction into partial fractions (after expanding the denominator to make things a bit more complicated), and then we factorise each element of the result with the help of **map**.

(c1) e1:(x+2)/((x+3)*(x+b)*(x-c)**2);

$$
\text{(d1)} \qquad \frac{x + 2}{(x + 3)\ (x + b)\ (x - c)^2}
$$

(c2) e1:ratsimp(e1);

$$
\text{(d2)} \qquad (x + 2)/(x^4 + (-\ 2\ c + b + 3)\ x^3
$$

$$+ (c^2 + (- 2\ b - 6)\ c + 3\ b)\ x^2 + ((b + 3)\ c^2 - 6\ b\ c)\ x$$

$$+ 3\ b\ c^2\ )$$

(c3) e2:partfrac(e1,x);

(d3) $- (c^2 + 4\ c - b + 6)/((c^4 + (2\ b + 6)\ c^3$

$$+ (b^2 + 12\ b + 9)\ c^2 + (6\ b^2 + 18\ b)\ c + 9\ b^2)\ (x - c))$$

$$+ \ \frac{c + 2}{(c^2 + (b + 3)\ c + 3\ b)\ (x - c)^2}$$

$$+ \ \frac{b - 2}{((b - 3)\ c^2 + (2\ b^2 - 6\ b)\ c + b^3 - 3\ b^2)\ (x + b)}$$

$$- \ \frac{1}{((b - 3)\ c^2 + (6\ b - 18)\ c + 9\ b - 27)\ (x + 3)}$$

(c4) e2:map(factor,e2);

$$(d4)\ -\ \frac{c^2 + 4\ c - b + 6}{(c + 3)^2\ (c + b)\ (x - c)}\ +\ \frac{c + 2}{(c + 3)\ (c + b)\ (x - c)^2}$$

$$+\ \frac{b - 2}{(b - 3)\ (c + b)^2\ (x + b)}\ -\ \frac{1}{(b - 3)\ (c + 3)^2\ (x + 3)}$$

For a given polynomial, `polydecomp` returns a list of polynomials which when composed together yield the input polynomial:

(c5) big:1769472*s**8 -3538944*s**7 +3907584*s**6-2764800*s**5

+1378560*s**4 -484608*s**3 +119520*s**2 -18768*s +1585$

(c6) polydecomp(big,s);

$$(d6) \qquad [3\ s^2 - 2,\ \frac{s^2 - 13}{12},\ 6\ s^2 + 11,\ 4\ s - 1]$$

Finally one can define simplification rules by the function **let** and apply them by the function **letsimp**.

```
(c1) let(a**2,%i);
```

$$(d2) \qquad a^2 \longrightarrow \%i$$

```
(c3) e0: a**2*(a**4-a)- a**3 +1;
```

$$(d3) \qquad a^2\ (a^4 - a) - a^3 + 1$$

```
(c4) letsimp(e0);
```

$$(d4) \qquad - 2\ \%i\ a - \%i + 1$$

```
(c5) let(om**3,1);
```

$$(d5) \qquad om^3 \longrightarrow 1$$

```
(c6) let(om**2, -1-om)$
```

```
(c7) pxyz: (x+y+z)*(x+om*y+om**2*z)*(x+om**2*y+om*z);
```

$$(d7) \qquad (z + y + x)\ (om\ z + om^2\ y + x)\ (om^2\ z + om\ y + x)$$

```
(c8) pxyz: letsimp(expand(%));
```

$$(d8) \qquad z^3 - 3\ x\ y\ z + y^3 + x^3$$

At the end of this section on simplification, we must point out the impact of programming style. By using certain LISP-inspired mechanisms available to the user, a considerable amount of time can sometimes be saved. We demonstrate this phenomenon by two functions **height1** and **height2** which, in FORTRAN style for the first and LISP for the second, find the greatest coefficient (in absolute value) of a polynomial over **Z**. The function **construct** is used to produce a polynomial of given degree and with integral coefficients chosen at random between -2^{35} and $2^{35}-1$. To help the reader understand the LISP form we have added some short examples which use the functions **numfactor**, **maplist** and **apply** directly. For more details the reader should consult the MACSYMA manual. The times are given in milliseconds for calculations done on the Honeywell computer DPS8-70M.

```
(c29) height1(poly,var):= block([m, maxi, i],
            maxi:0,
            for i:0 thru hipow(poly,var) do
                ( m: abs(coeff(poly,var,i)),
                    if m>maxi then maxi:m),
            return(maxi) )$
```

```
(c30) height2(poly):= apply(max,maplist(absfact,poly))$
```

```
(c31) absfact(z):= abs(numfactor(z)) $
```

```
(c32) construct (n):= block([pol,i],
            pol:0,
            for i:1 thru n do pol:pol+random() * x^i,
            return (pol) );
```

```
(c33) p0:construct(10);
```

$$(d33) \quad -13379176882 \ x^{10} + 8895517289 \ x^{9} - 9101752241 \ x^{8} - 426691508 \ x^{7}$$

$$+ 17446139911 \ x^{6} + 33598445718 \ x^{5} + 33193902370 \ x^{4} + 3493507944 \ x^{3}$$

$$- 5008989035 \ x^{2} - 17002352646 \ x$$

```
(c34) numfactor( -34*x**2);
```

$$(d34) \qquad\qquad -34$$

```
(c35) maplist(absfact,p0);
```

(d35) [1337917688, 8895517289, 9101752241, 426691508, 17446139911, 33598445718, 33193902370, 3493507944, 5008989035, 17002352646]

```
(c36) apply(max,%);
```

$$(d36) \qquad\qquad 33598445718$$

```
(c37) p1:construct (20)$
```

```
(c38) height1(p1,x);
```

    Time= 851 msec.

$$(d38) \qquad\qquad 33598445718$$

```
(c39) height2(p1);
```

    Time= 272 msec.

$$(d39) \qquad\qquad 33598445718$$

```
(c40)  p2:construct (40)$

(c41)  height1(p2,x);

      Time= 2268 msec.

(d41)                          32727645108

(c42)  height2(p2);

      Time= 529 msec.

(d42)                          32727645108

(c43)  p100:construct(100)$

(c44)  height1(p100,x);

      Time= 12030 msec.

(d44)                          34073350797

(c45)  height2(p100);

      Time= 1577 msec.

(d45)                          34073350797
```

1.5.6 Integration

The mechanisation of the search for integrals has always been an important part of Computer Algebra, because of the prestige attaching to the problem. From a practical point of view it is more important for a Computer Algebra system to be efficient in the treatment of rational expressions than in looking for integrals. But this problem has led to great progress in this aspect of machine use. The names Slagle, Moses, Risch, are closely linked with this. MACSYMA has been one of the first systems to benefit from an advanced integration command. REDUCE benefited subsequently from the latest improvements following the work of the Cambridge (G.B.) group.

The command in MACSYMA is **integrate**. The results appear in a form which may need rearranging (for example the result d2).

```
(c1)  integrate(1/(1-x**4),x);
```

$$(d1) \quad \frac{\log(x + 1)}{4} + \frac{\operatorname{atan}(x)}{2} - \frac{\log(x - 1)}{4}$$

```
(c2)  integrate(sinh(x)**4,x);
```

```
       4 x
      %e                 2 x           - 2 x     %e
      -----  - 2 %e        + 2 %e          - ------- + 6 x
        4                                      4
(d2)  ----------------------------------------------------
                             16
```

(c3) `integrate(x**3*cos(x**2),x);`

```
               2      2       2
              x  sin(x) + cos(x)
(d3)          -------------------
                       2
```

Sometimes the final form of the result may be surprising:

(c4) `integrate(sin(a/x)-cos(x**2),x);`

(d4) `(sqrt(%pi) ((sqrt(2) %i - sqrt(2))`

```
       (sqrt(2) %i + sqrt(2)) x
 erf(------------------------)
               2
```

```
                             (sqrt(2) %i - sqrt(2)) x
  + (sqrt(2) %i + sqrt(2)) erf(------------------------))
                                       2
```

```
                      a
              / cos(-)
        a     [     x
  + 8 sin(-) x + 8 a I ------ dx)/8
        x     ]     x
              /
```

Definite integration is also possible. The symbol `inf` is the symbol MACSYMA reserves for ∞. If the command **integrate** needs information, a question is put to the user.

(c5) `integrate(1/sqrt((1+x)**3 * (1+ 2*x)),x, 0,inf);`

(d5)   `2 sqrt(2) - 2`

(c6) `integrate(1/sqrt((1+x)**3 * (1 + k*x)),x,0,inf);`

`Is  k  positive, negative, or zero?`

`positive;`

`Is  k - 1  zero or nonzero?`

`nonzero;`

Is k - 1 positive or negative?

positive;

(d6)
$$\frac{2\ \text{sqrt}(k)}{k - 1} - \frac{2}{k - 1}$$

(c7) integrate(1/sqrt(x * (u-x)),x,0,u);

Is u positive, negative, or zero?

positive;

(d7)   %pi

The Laplace transform with the two commands **laplace** and **ilt** (the inverse) should be mentioned together with definite integration.

(c12) laplace(%e**(a*t)*(b*t**2+c*t-sin(d*t)),t,s);

$$(d12) \ - ((d - c)\ s^3 + (- 3\ a\ d + 3\ a\ c - 2\ b)\ s^2$$

$$+ (- c\ d^2 + 3\ a^2\ d - 3\ a^2\ c + 4\ a\ b)\ s + (a\ c - 2\ b)\ d^2$$

$$- a^3\ d + a^3\ c - 2\ a^2\ b)/(s^5 - 5\ a\ s^4 + (d^2 + 10\ a^2)\ s^3$$

$$+ (- 3\ a\ d^2 - 10\ a^3)\ s^2 + (3\ a^2\ d^2 + 5\ a^4)\ s - a^3\ d^2 - a^5)$$

(c13) ilt(%,s,u);

Is d   zero or nonzero?

nonzero;

$$(d13) \quad - \%e^{a\ u}\ \sin(d\ u) + b\ u^2\ \%e^{a\ u} + c\ u\ \%e^{a\ u}$$

The Laplace transform can be used to solve an integral equation. In *c*15 we see the use of the symbol ' which prevents evaluation. This is similar to the **quote** of Lisp. The result *d*15 shows it explicitly. Similarly the Laplace transform of an unknown function remains unevaluated.

```
(c15) 'integrate(cosh(a*x)*f(t-x),x,0,t) +b*f(t)=t**3;

          t
         /
         [                                          3
(d15)    I   f(t - x) cosh(a x) dx + b f(t) = t
         ]
         /
          0

(c16) laplace(%,t,s);

                             s laplace(f(t), t, s)     6
(d16) b laplace(f(t), t, s) + --------------------- = --
                                     2     2            4
                                    s   - a            s
```

We solve this linear equation for $laplace(f(t), t, s)$

```
(c17) linsolve([%], ['laplace(f(t),t,s)]);

                                 2       2
                              6 s   - 6 a
(d17)    [laplace(f(t), t, s) = --------------------]
                                 6     5     2     4
                                b s  + s  - a  b s

(c18) formu:% $
```

Now we have to extract the interesting part, that is the right hand side of the equation d17. For that we use the function **part** with arguments linked to the structure of the formula. But we can check in advance that we do extract the desired part, thanks to the function **dpart**.

```
(c19) dpart(formu[1],2);

                          """""""""""""""""""""""""""""""""
                          "        2       2            "
                          "     6 s   - 6 a             "
(d19)    laplace(f(t), t, s) = "--------------------"
                          "     6     5     2     4"
                          "b s  + s  - a  b s "
                          """""""""""""""""""""""""""""""""

(c20)  res: part(formu[1],2)$

(c21)  ilt(res,s,t);
```

```
           t
      -  ---             2  2             2  2
         2 b        6 a  b   + 6     2 (12 a  b   + 6)
(d21) %e           ((------------  -  ------------------)
                        6  3              6  3
                       a  b              a  b

           2  2
   sqrt(4 a  b   + 1) t                  2  2
sinh(---------------------)/sqrt(4 a  b   + 1)
          2 b

                              2  2
     2  2               sqrt(4 a  b   + 1) t
 (6 a  b   + 6) cosh(---------------------)            3       2
                          2 b                   t     3 t
- --------------------------------------------)/b + -- + -----
              6  3                                  b    2  2
             a  b                                       a  b

            2  2
   6 t     6 a  b   + 6
+ ----- + -----------
   4  3        6  4
  a  b        a  b
```

We conclude with a simple example of the solution of a Fredholm integral equation of the second type. We want to solve the equation

$$f(x, \lambda) = x + \lambda x^3 \int_0^1 y f(y, \lambda) dy + \lambda x^2 \int_0^1 y^2 f(y, \lambda) dy + \lambda x \int_0^1 y^3 f(y, \lambda) dy \tag{1}$$

by the following method:
   (1) We put

$$A(\lambda) = \int_0^1 y f(y, \lambda) dy$$

$$B(\lambda) = \int_0^1 y^2 f(y, \lambda) dy$$

$$C(\lambda) = \int_0^1 y^3 f(y, \lambda) dy$$

Equation (1) becomes

$$f(x, \lambda) = x + \lambda x^3 A(\lambda) + \lambda x^2 B(\lambda) + \lambda x C(\lambda) \tag{2}$$

(2) In the definitions of A, B and C, we replace $f(y, \lambda)$ by its value from equation (2) with x replaced by y. If we integrate, we get 3 linear equations in the 3 unknowns A, B and C.

(3) We solve this linear system and introduce the values found into (2). Thus we get the desired function.

Solving it with the help of MACSYMA closely follows the algorithm. To simplify the input, we start by defining a useful function $f(q1, q2)$.

```
(c2) f(q1,q2):= q2*a*q1^3+q2*b*q1^2+q2*c*q1+q1;
```

$$\text{(d2)} \quad f(q1, q2) := q2 \ a \ q1^3 + q2 \ b \ q1^2 + q2 \ c \ q1 + q1$$

```
(c3) eqa: a=integrate(y*f(y,1),y,0,1);
```

$$\text{(d3)} \qquad a = \frac{(20 \ c + 15 \ b + 12 \ a) \ 1 + 20}{60}$$

```
(c4) eqb: b=integrate(y^2*f(y,1),y,0,1);
```

$$\text{(d4)} \qquad b = \frac{(15 \ c + 12 \ b + 10 \ a) \ 1 + 15}{60}$$

```
(c5) eqc: c=integrate(y^3*f(y,1),y,0,1);
```

$$\text{(d5)} \qquad c = \frac{(42 \ c + 35 \ b + 30 \ a) \ 1 + 42}{210}$$

```
(c6) sol: solve([eqa,eqb,eqc],[a,b,c]);
```

$$\text{(d6)} \quad [[a = - \frac{1575 \ 1 - 126000}{1^3 - 4140 \ 1^2 - 226800 \ 1 + 378000},$$

$$b = \frac{2100 \ 1 + 94500}{1^3 - 4140 \ 1^2 - 226800 \ 1 + 378000},$$

$$c = - \frac{1^2 - 3510 \ 1 - 75600}{1^3 - 4140 \ 1^2 - 226800 \ 1 + 378000}]]$$

We now extract the useful values of the result *vector*, after checking a by dpart.

(c7) dpart(sol,1,1,2);

```
             """""""""""""""""""""""""""""""""""""""""""""""""""""
             "            1575 l - 126000               "
(d7) [[a = "-  ------------------------------------",
             "    3         2                          "
             "   l  - 4140 l  - 226800 l + 378000"
             """""""""""""""""""""""""""""""""""""""""""""""""""""

                      2100 l + 94500
        b =    -------------------------------------,
                 3         2
                l  - 4140 l  - 226800 l + 378000

                      2
                     l  - 3510 l - 75600
        c = - ------------------------------------]]
                 3         2
                l  - 4140 l  - 226800 l + 378000
```

(c8) a: part(sol,1,1,2) $

(c9) b: part(sol,1,2,2) $

(c10) c: part(sol,1,3,2) $;

(c11) ratdenomdivide:false$

Now we have to evaluate $f(x,l)$ by replacing a, b and c by their value. We can use **ev** or a variant as is seen in *c*12.

(c12) f(x,l),eval;

```
                                          3
              l (1575 l - 126000) x
(d12)  -  ------------------------------------
              3         2
             l  - 4140 l  - 226800 l + 378000

                                        2
              l (2100 l + 94500) x
     +  -----------------------------------
          3         2
         l  - 4140 l  - 226800 l + 378000

           2
          l (l  - 3510 l - 75600) x
     -  ------------------------------------ + x
          3         2
         l  - 4140 l  - 226800 l + 378000
```

Finally we arrange and simplify the result using **ratvars** and **ratsimp**. For the sake of a later example we also substitute $l = 1$.

```
(c13) ratvars(x,1);

(d13)                        [x, 1]

(c14) ratsimp(d12);

           2      3           2
(d14) - (1   (1575 x  - 2100 x  + 630 x)

            3           2
   + 1 (- 126000 x  - 94500 x  + 151200 x) - 378000 x)

    3        2
 /(1  - 4140 1  - 226800 1 + 378000)


(c15) subst(1,1,%);

                 3          2
          124425 x    96600 x    226170 x
(d15)     --------- + -------- + --------
           147061      147061     147061
```

1.6 AVAILABILITY OF MACSYMA

The preceding examples have given only a brief survey of MACSYMA's possibilities. The interested reader should practise on a real system. He might also consider aspects which we have not described here: functions for drawing curves and surfaces, and especially the huge possibilities of the library SHARE. The best versions of MACSYMA are installed on VAX computers, on the LISP machines of the firms SYMBOLICS and LMI, and on SUN work-stations.

1.7 OTHER SYSTEMS

The present supremacy of MACSYMA's library should not make us over-look the existence of other systems for Computer Algebra. REDUCE, which is described later (see the Annex), has developed very rapidly over the last few years. It is to be found on almost all machines in the scientific world. For its execution LISP uses as pre-processor RLISP, a system which is easy to understand and to adapt. We should also mention MAPLE (written in the language C and available under UNIX) which is fairly wide-spread, as are SMP and AMP (written in assembly language for IBM mainframes). As a starting system we must mention muMATH which works on micro-computers. As an example of its use we go back to the integral equation we looked at earlier.

As in MACSYMA, we can define an auxiliary function F. Note the double call to EVSUB to get FY.

```
? F:Q2*A*Q1^3+Q2*B*Q1^2+Q2*C*Q1+Q1 ;

@: Q1 + A Q2 Q1^3 + B Q2 Q1^2 + C Q2 Q1

? FY:EVSUB(EVSUB(F,Q1,Y),Q2,L) ;

@: Y + A Y^3 L + B Y^2 L + C Y L
```

We define the 3 equations in which there is an integral.

```
? EQA:DEFINT(Y*FY,Y,0,1) == A ;

@: 1/3 + A L/5 + B L/4 + C L/3 == A

? EQB:DEFINT(Y^2*FY,Y,0,1) == B ;

@: 1/4 + A L/6 + B L/5 + C L/4 == B

? EQC:DEFINT(Y^3*FY,Y,0,1) == C ;

@: 1/5 + A L/7 + B L/6 + C L/5 == C
```

Since muMATH does not have a function like the **solve** of MACSYMA, we construct *manually* the matrix MA of the linear system with A, B and C as the unknowns. When we have extracted the second member MB in the same way, we solve $MA.V = MB$.

```
? MA:{[L/5-1,L/4,L/3],[L/6,L/5-1,L/4],[L/7,L/6,L/5-1]};

@: {[(-5 + L)/5, L/4, L/3],
    [L/6, (-5 + L)/5, L/4],
    [L/7, L/6, (-5 + L)/5]}

? MB:{[-1/3],[-1/4],[-1/5]} ;

@: {[-1/3],
    [-1/4],
    [-1/5]}

? V:MA \ MB $
```

By putting $L = 1$ in the solution V, and by evaluating we rediscover the coefficients of the final polynomial of the example in MACSYMA.

```
? L:1 $
```

```
?  EVAL(V)  ;

@:  {[124425/147061],
    [96600/147061],
    [79109/147061]}
```

Finally we cannot end without mentioning the latest system of Computer Algebra: SCRATCHPAD. This system is written in the LISP of IBM's system VM-CMS. Although it has not reached the degree of development of REDUCE or of MACSYMA in the matter of commands and of the libraries available for the solution of ordinary problems, it stands out from these *classic* systems by its new approach which is likely to appeal especially to mathematicians. Let us go back for a moment to MACSYMA. It is true that this system has some ways of working with algebraic numbers. But the results are obtained at the expense of internal transformations between the *recursive* form and the *tree* form, and we shall return to this later. This is the reason for the function simplify in the examples which follow.

```
(c1)  algebraic:true;

(d1)                              true

(c2)  simplify(e):=ratdisrep(rat(e));

(d2)          simplify(e) := ratdisrep(rat(e))

(c3)  (5*(2-%i))/(3^(1/3)+%i);

                            5 (2 - %i)
(d3)                        ----------
                                  1/3
                            %i + 3

(c4)  simplify(%);

            2/3      1/3                  2/3    1/3
           (3    + 7 3    - 1) %i - 7 3    + 3    + 7
(d4)    -  -------------------------------------------
                                  2
```

If we assign a prime number p as the value of the global variable modulus, we are working over the field $\mathbf{Z}_p$, with a representation of the elements which is symmetric with respect to 0 (including $\mathbf{Z}_2$!). For example:

$$Z_7 = (-3, -2, -1, 0, 1, 2, 3)$$

The default value of **modulus** is **false**. We then work in **Z**, **Q** or **R** as the case may be, and as we have done for all the examples in this chapter.

(c5) modulus:7;

(d5)   7

(c6) p0: x^16+4;

$$(d6) \qquad x^{16} + 4$$

(c7) factor(p0);

$$(d7)\ (x^4 - 2 x^2 + 3)\ (x^4 - x^2 + 3)\ (x^4 + x^2 + 3)$$

$$(x^4 + 2 x^2 + 3)$$

(c8) ratexpand(%);

$$(d8) \qquad x^{16} - 3$$

(c9) modulus:2;

(d9)   2

(c10) factor((x^16-x)/(x^4-x));

$$(d10)\ (x^4 + x + 1)\ (x^4 + x^3 - 1)\ (x^4 + x^3 + x^2 + x + 1)$$

(c11) simplify((x^16-x)/(x^4-x));

$$(d11) \qquad x^{12} + x^9 + x^6 + x^3 + 1$$

(c12) factor(%);

$$(d12)\ (x^4 + x + 1)\ (x^4 + x^3 - 1)\ (x^4 + x^3 + x^2 + x + 1)$$

We can also define extensions of the ground field. That is the role of the function **tellrat**. There is no check for irreducibility of the defining polynomials.

(c13) tellrat(1+a+a^2);

$$(d13) \qquad [a^2 + a + 1]$$

(c14) z0: 1+a+a^3+a^5+ 1/a;

$$(\text{d}14) \qquad a^5 + a^3 + a + \frac{1}{a} + 1$$

(c15) z0: simplify(z0);

(d15) a

(c16) simplify((x^2+x+a)*(x^2+x+a^2));

$$(\text{d}16) \qquad x^4 + x + 1$$

(c17) tellrat(b^2+b+a);

$$(\text{d}17) \qquad [b^2 + b + a, \ a^2 + a + 1]$$

(c18) z1:a+b;

(d18) b + a

(c19) iz1: simplify(1/z1);

(d19) a b + 1

(c20) simplify(z1*iz1);

(d20) 1

(c21) tellrat(g^2+b*g+a+1);

$$(\text{d}21) \qquad [g^2 + b\,g + a + 1, \ b^2 + b + a, \ a^2 + a + 1]$$

(c22) z2: a+b+g+1;

(d22) g + b + a + 1

(c23) iz2: simplify(1/z2);

(d23) ((a + 1) b + a) g + a b + 1

Let us go back to **Q**. The constant $z2$ always has formally the value $1 + a + b + g$.

(c24) modulus:false;

(d24) false

```
(c25) simplify(1/z2);

        ((a + 3) b - 3 a - 2) g + (- 3 a - 2) b + 2 a - 1
(d25) - ----------------------------------------------------
                                7

(c26) simplify(1/b);

(d26)                       (a + 1) b + a + 1

(c27) a0: a+6*b*g;

(d27)                           6 b g + a

(c28) a1: simplify(1/a0);

(d28) - (((52422 a + 242820) b + 60084 a + 247536) g

        + (17676 a + 45972) b -- 212117 a + 102415)/1244803

(c29) simplify(1/a1);

(d29)                           6 b g + a
```

Unfortunately, as we shall see later, MACSYMA has some problems when factorising the results in these extensions and even in simplifying the rational fractions by taking the g.c.d.

```
(c30) p0: x^2 + 3*b*x +a*g+2;

                              2
(d30)                       x  + 3 b x + a g + 2

(c31) p1: (x-a)*(x-g);

(d31)                       (x - a) (x - g)

(c32) p3: simplify(p0*p1);

      4                        3
(d32) x  + (- g + 3 b - a) x  + ((2 a - 3 b) g - 3 a b + 2)

  2
 x  + ((4 a b + a - 1) g - 2 a - 1) x + ((a + 1) b + 2 a) g

 + a

(c33) simplify(p3/p1);
```

(d33) $(x^4 + (-g + 3b - a) x^3$

$+ ((2a - 3b) g - 3ab + 2) x^2$

$+ ((4ab + a - 1) g - 2a - 1) x + ((a + 1) b + 2a) g$

$+ a)/(x^2 + (-g - a) x + a g)$

(c34) factor(p3);

(d34) $x^4 - g x^3 + 3 b x^3 - a x^3 - 3 b g x^2 + 2 a g x^2$

$- 3 a b x^2 + 2 x^2 + 4 a b g x + a g x - g x - 2 a x - x$

$+ a b g + b g + 2 a g + a$

Even if useful, these MACSYMA applications are incomplete and demand a certain knowledge of the internal working of the system. The user may be aware of an unnatural aspect which easily degenerates into hacking. It is this that the authors of SCRATCHPAD were trying to avoid and so they thought in terms of mathematical structure and of *generic* operations. A full study would go beyond the limits which the authors of this book have set themselves. The reader should refer to the documentation on SCRATCHPAD which at the moment is very brief but which will certainly grow with our knowledge of SCRATCHPAD, as it develops through lectures and courses. But here are some small examples which give an idea of how SCRATCHPAD is used. We want to write a function `quotient` which allows us to divide two polynomials according to *increasing* powers of the variable. First we have to define the type of function, that is that its first two arguments are in the domain of the polynomials x over the rationals $(UP(x, RN))$, the third comes from the domain of non-negative integers (NNI) and the result is in the domain $UP(x, RN)$ or the symbol "failed". The real definition of the function `quotient` uses primitive operations which are similar to those used in MACSYMA. It is important to note also the type declarations of each variable before using it.

```
quotient:(UP(x,RN),UP(x,RN),NNI) -> Union("failed", UP(x,RN))
```

```
quotient(p1,p2,n) ==
      mindeg(p1) < mindeg(p2) => "failed"
      reste:UP(x,RN):=p1
      degp2:mindeg(p2)
      coefp2:coef(p2,degp2)
      quot:UP(x,RN):=0
      while degree(quot) < n  repeat
            deg:=mindeg(reste)
            mon:=monom(deg-degp2,coef(reste,deg)/coefp2)
            num:=num-(mon*p2)
            quot:=quot+mon
      quot

 0  total errors
```

This function can now be used, but the SCRATCHPAD system compiles it the first time it is called.

quotient(1,1+x,8)

```
   compiling quotient with signature
   (P[x]RN,P[x]RN,NNI) -> Union(failed,P[x]RN)

       8    7    6    5    4    3    2
   (3)  x  - x  + x  - x  + x  - x  + x  - x + 1
```

Type: Union(failed,P[x]RN)

quotient(x**2-x+1,x**3-x-6/7,8)

```
   (5)
    84778967    8    - 18089477    7     2286095    6     - 166061    5
   (----------)x  + (------------)x  + (---------)x  + (----------)x
    10077696          1679616           279936           46656


        - 8281   4     4459   3     - 889   2     91         - 7
   +   (--------)x  + (------)x  + (-------)x  + (----)x + -----
        7776          1296          216          36          6
```

Type: Union(failed,P[x]RN)

quotient(x**2-x+1,x**3-x-6/7,8)

```
   (6)

    84778967    8    - 18089477    7     2286095    6     - 166061    5
   (----------)x  + (------------)x  + (---------)x  + (----------)x
    10077696          1679616           279936           46656


        - 8281   4     4459   3     - 889   2     91         - 7
   +   (--------)x  + (------)x  + (-------)x  + (----)x + -----
        7776          1296          216          36          6
```

Let us now take an example of an algebraic extension, for which MAC-SYMA was not completely satisfactory. We use the field `SimpleAlgebraicExtension` by taking a field, a polynomial structure and a polynomial in parameter. That lets us define a new type: `ext1`.

```
ext1:=SAE(RN,UP(a,RN),a**2+a+1)
```

```
loading SAE LISPLIB C for domain SimpleAlgebraicExtension
loading MM LISPLIB C for domain ModMonic
```

```
   (7)  SAE(RN,P[a]RN,(a**2)+a+1)
```

Type: aCategory

```
e:ext1:=(3/4)*a**2-a+(7/4)
```

```
loading DMP LISPLIB C for domain DistributedMultivariatePolynomial
loading DP LISPLIB C for domain DirectProduct
loading OV LISPLIB C for domain OrderedVarlist
loading MP LISPLIB C for domain MultivariatePolynomial
loading RF LISPLIB C for domain RationalFunction
loading NRATINT LISPLIB C for package NuRatint
loading IR LISPLIB C for domain IntegrationResult
```

$$
(8) \quad \left(\frac{-7}{4}\right)a + 1
$$

Type: SAE(RN,P[a]RN,(a**2)+a+1)

```
-- inversion of e:
```

```
recip$ext1 (e)
```

$$
(9) \quad \left(\frac{28}{93}\right)a + \frac{44}{93}
$$

Type: Union(SAE(RN,P[a]RN,(a**2)+a+1),failed)

```
e**2
```

$$
(10) \quad \left(\frac{-105}{16}\right)a + \frac{-33}{16}
$$

Type: SAE(RN,P[a]RN,(a**2)+a+1)

```
e:=a**2-1
```

$$
(11) \quad - a - 2
$$

Type: SAE(RN,P[a]RN,(a**2)+a+1)

We take two expanded polynomials *p1* and *p2*, with coefficients in this extension equal respectively to $(x^2 - a^2)(x^2 + 3x + a)$ and $x^2 - a^2$. This time we see that the system simplifies correctly the rational fraction *p1/p2*.

```
p1:UP(x,ext1):=x**4+3*x**3+(2*a+1)*x**2+(3*a+3)*x-1
```

$$(12) \quad x^4 + 3 x^3 + (2 a + 1)x^2 + (3 a + 3) x -1$$

```
Type: P[x]SAE(RN,P[a]RN,(a**2)+a+1)
```

```
p2:UP(x,ext1):= x**2+a+1
```

$$(13) \quad x^2 + a + 1$$

```
Type: P[x]SAE(RN,P[a]RN,(a**2)+a+1)
```

```
p2/p1
```

$$(14) \quad x^2 + 3x + a$$

```
Type: P[x]SAE(RN,P[a]RN,(a**2)+a+1)
```

As the latest of the Computer Algebra systems offered to mathematicians and engineers SCRATCHPAD should certainly have a considerable impact. As with the others, it is the use of it which will lead to its development and it is to be hoped that its authors will make it available to everyone at as many sites as possible.

2. The problem of data representation

This chapter is devoted to a basic aspect of Computer Algebra: the representation of mathematical objects on the computer. In this chapter we shall consider questions of the form: "How are the data represented on the computer?"; "I can present my problem in several ways — which will be the most efficient?" etc. We shall not go into the very technical questions of REDUCE or MACSYMA, but we shall explain the general principles, and the representation traps into which the user may so easily fall.

2.1 REPRESENTATIONS OF INTEGERS

"God created the integers: the rest is the work of man" said the great mathematician Kronecker. Nevertheless, man has some problems with representing integers. Most computer languages treat them as a finite set, such as $\{-2^{31}, \ldots, 2^{31} - 1\}$. It is very easy to say "my data are small, the answer will be small, so I shall not need large integers and this finite set will be enough", but this argument is completely false.

Suppose we want to calculate the g.c.d. of the following two polynomials [Knuth, 1969; Brown, 1971]:

$$A(x) = x^8 + x^6 - 3x^4 - 3x^3 + 8x^2 + 2x - 5;$$
$$B(x) = 3x^6 + 5x^4 - 4x^2 - 9x + 21.$$

Since we know Euclid's algorithm we can begin to calculate (we shall return to this example in the sub-section "The g.c.d."). The last integer calculated has 35 decimal digits, that is 117 bits, and as the result is an integer we find that the polynomials are relatively prime. However, the data are small (the biggest number is 21), and the result "relatively prime" was a yes/no answer and only needed 1 bit. This problem of *intermediate expression swell* is one of the biggest problems in Computer Algebra. Similar problems have been pointed out in integration [Davenport, 1981, pp. 165–172] and elsewhere. In the chapter "Advanced algorithms" we give an example of an algorithm for the g.c.d. of polynomials, which does not have this defect.

So, it is necessary to deal with the true integers of mathematicians, whatever their size. Almost all Computer Algebra systems do this, and a trivial test of a Computer Algebra system is to calculate such a number. The representation of large integers (sometimes called "bignums") is fairly obvious: we choose a number N as base, just as the normal decimal representation uses 10 as base, and every number is represented by its sign (+ or −) and a succession of "digits" (that is the integers between 0 and $N - 1$) in this base. The most usual bases are the powers of 2 or of 10: powers of 10 make the input and the output of the numbers (that is their conversion from or into a decimal representation) easier, whilst powers of 2 make several internal calculations more efficient, since multiplication by the base and division by the base can be done by a simple shift. This question is studied in Davenport and Padget [1987]. Normally it is worth choosing as large a base as possible (provided that the "digits" can be stored in one word). For example, on a computer with 32 bits (such as the large IBM, 43XX and 30XX, and the Motorola 68020), we can choose 10^9 as the decimal base, 2^{30} or 2^{31} as the binary base. Usually 2^{32} is not used, because it makes addition of numbers difficult: the carry (if there is one) cannot be stored in the word, but has to be retrieved in a way which is completely machine-dependent ("overflow indicator" etc.).

Once we have chosen such a representation, addition, subtraction and multiplication of these integers are, in principle at least, fairly easy — the same principles one learns in primary school for decimal numbers suffice. Nevertheless, the multiplication of numbers gives rise to a problem. The product of two numbers each contained in one word requires two words for storing it. Most computers contain an instruction to do this, but "high level" languages do not give access to this instruction (the only exception is BCPL [Richards and Whitby-Strevens, 1979], with its MULDIV operation). Therefore almost all the systems for large integers contain a function written in machine language to do this calculation.

Division is much more complicated, since the method learnt at school is not an algorithm as it requires us to guess a number for the quotient. Knuth [1981] describes this problem in detail, and gives an algorithm for this guesswork, which is almost always right, can never give too small an answer and can never give an answer more than one unit too big (in such an event, the divisor has to be added to the numerator to correct this error). To calculate the g.c.d. there is Euclid's algorithm and also several other ways which may be more efficient. Knuth [1981] treats this problem too.

But the fact that the systems *can* deal with large numbers does not mean that we *should* let the numbers increase without doing anything. If we have two numbers with n digits, adding them requires a time proportional to n, or in more formal language a time $O(n)$. Multiplying them requires a time $0(n^2)$*. Calculating a g.c.d., which is fundamental in the calculation of rational numbers, requires $O(n^3)$, or $O(n^2)$ with a bit of care**. This implies that if the numbers become 10 times longer, the time is multiplied by 10, or by 100, or by 1000. So it is always worth reducing the size of these integers. For example it is more efficient to integrate $1/x + 2/x^2$ than to integrate $12371265/(249912457x) + 24742530/(249912457x^2)$.

When it comes to factorising, the position is much less obvious. The simple algorithm we all know, which consists of trying all the primes less than $N^{1/2}$, requires a running time $O(N^{1/2}\log^2 N)$, where the factor of $\log^2 N$ comes from the size of the integers involved. If N is an integer with n digits, it becomes $O(10^{n/2}n^2)$, which is an exponential time. Although they are not at present† implemented in any Computer Algebra system which is widely used, better algorithms do exist with a running time which increases more slowly than the exponentials, but more quickly than the polynomials, that is $O(\exp((n\log n)^{1/2}))$. For integers of a realistic size, we find [Wunderlicht, 1979] $O(N^{.154})$, which is a little less than $O(10^{n/6})$. Macmillan and Davenport [1984] give a brief account of what is meant by "realistic". Much research is being done on this problem at present, largely because of its cryptographic interest. In general, we can say that the reflex "I have just calculated an integer; now I am going to factorise it" is very dangerous.

2.2 REPRESENTATIONS OF FRACTIONS

Fractions (that is the rational numbers) are obviously almost always represented by their numerator and denominator. As a rule they must not be replaced by floating numbers for this entails not only loss of precision, but completely false results‡. For example, the g.c.d. of $x^3 - 8$

---

* In principle, $O(n\log n \log\log n)$ is enough [Aho *et al.*, 1974, Chapter 8], but no computer algebra system uses this at the present time, for it is more like $20n\log n\log\log n$.

** In principle, $O(n\log^2 n\log\log n)$ [Aho *et al.*, 1974, Chapter 8], but again no system uses it.

† This was true when the book was written. While the translation was being completed, REDUCE version 3.3 was released, which possesses an advanced factoring algorithm.

‡ For example, the system SMP presents several problems arising from the use of such a representation for rational numbers.

and $(1/3)x^2 - (4/3)$ is $(1/3)x - (2/3)$, whereas the g.c.d. of $x^3 - 8$ and $.333333x^2 - 1.33333$ is $.000001$, because of truncation.

All calculations with rational numbers involve calculating g.c.d.s, and this can be very costly in time. Therefore, if we can avoid them, the calculation will usually go much better. For example, instead of the above calculation, we can calculate the g.c.d. of $x^3 - 8$ and $x^2 - 4$, and this can be calculated without any g.c.d. over the integers. The same is true of the relation polynomials — rational functions, and we shall show an algorithm (Bareiss' algorithm) which does not need recourse to fractions for elimination in matrices.

The algorithms of addition, multiplication etc. of fractions are fairly easy, but nevertheless there are possible improvements. Let us consider for example multiplication:

$$\frac{a}{b} \times \frac{c}{d} = \frac{p}{q}.$$

The most obvious way is to calculate $p = ac$, $q = bd$, and then to remove the g.c.d. But, supposing that a/b and c/d are already reduced, we can conclude that

$$\gcd(p, q) = \gcd(a, d)\gcd(b, c).$$

Then it is more efficient to calculate the two g.c.d.s on the right than the g.c.d. on the left, for the data are smaller.

It is the same for addition:

$$\frac{a}{b} + \frac{c}{d} = \frac{p}{q}.$$

Here we can calculate $p = ad + bc$ and $q = bd$, but it is more efficient to calculate $q = bd/\gcd(b, d)$ and

$$p = a\frac{d}{\gcd(b, d)} + c\frac{b}{\gcd(b, d)}.$$

It is still necessary to take out the g.c.d. of p and q, but this method yields p and q smaller than does the simpler method.

2.3 REPRESENTATIONS OF POLYNOMIALS

Now we shall consider the fundamental calculation of all Computer Algebra systems, that which distinguishes them from other systems: polynomial calculation. We must stress that the adjective "polynomial" applies to programmed calculations and not necessarily to the types of data to which the calculations apply. For example, the calculation

$$(x - y)(x + y) = x^2 - y^2$$

is a polynomial calculation, but

$$(\cos a - \sin b)(\cos a + \sin b) = \cos^2 a - \sin^2 b$$

is also one: in fact we have here the same calculation, with the variable x replaced by $\cos a$ and the variable y replaced by $\sin b$.

All Computer Algebra systems manipulate polynomials in several variables (normally an indefinite number*). We can add, subtract, multiply and divide them (at least if the division is without remainder — see later), but in fact the interesting calculation is *simplification*. It is not very interesting to write the product of $(x + 1)$ and $(x - 1)$ as $(x + 1)(x - 1)$: to write it as $x^2 - 1$ is much more useful.

"Simplification" is a word with many meanings. It is obvious that $x - 1$ is "simpler" than $(x^2 - 1)/(x + 1)$, but is $x^{999} - x^{998} + x^{997} - \cdots - 1$ "simpler" than $(x^{1000} - 1)/(x + 1)$? To deal with this somewhat nebulous idea, let us define exactly two related ideas which we shall need.

2.3.1 Canonical and normal representations

A representation of mathematical objects (polynomials, in the present case, but the definition is more general) is called *canonical* if two different representations always correspond to two different objects. In more formal terms, we say a correspondence f between a class O of objects and a class R of representations is *a representation* of O by R if each element of O corresponds to one or more elements of R (otherwise it is not represented) and each element of R corresponds to one and only one element of O (otherwise we do not know which element of O is represented). The representation is canonical if f is bijective. Thus we can decide whether two elements of O are equal by verifying that their representations are equal.

If O has the structure of a monoid (and almost every class in Computer Algebra is at least a monoid), we can define another concept. A representation is called *normal* if zero has only one representation. (One may ask "Why zero?" The reason is that zero is not legitimate as a second parameter for division, and therefore one has to test before dividing. If there were other excluded values, the definition of "normal" would be more complicated.) Every canonical representation is normal, but the converse is false (as we shall see later). A normal representation over a group gives us an algorithm to determine whether two elements a and b of O are equal: it is sufficient to see whether $a - b$ is zero or not. This algorithm is, of course, less efficient than that for canonical representations, where it is sufficient to see whether the representations of a and b are equal.

---

* CAMAL requires their number to be declared, but this constraint is the reason for the speed of CAMAL.

So we can say that a simplification should yield a representation which would be at least normal, and if possible canonical. But we want much more, and it is now that subjectivity comes into play. We want the simplified representation to be "regular", and this would exclude a representation such as the following (where we have defined $A = x^2 + x$, and we then demand some variations of A) :

object	:	representation
A	:	$x(x+1)$
$A+1$	:	$\dfrac{x^3 - 1}{x - 1}$
$A - x$	:	x^2
$A + x + 1$	:	$(x+1)^2$

Brown's storage method [1969] raises this question of regularity. Given a normal representation, he proposes to construct a canonical representation. All the expressions which have already been calculated are stored, $a_1, \ldots, a_n$. When a new expression b is calculated, for all the a_i we test whether b is equal to this a_i (by testing whether $a_i - b$ is zero). If $b = a_i$, we replace b by a_i, otherwise a_{n+1} becomes b, which is a new expression. This method of storing yields a canonical representation, which, however, is not at all regular, because it depends entirely on the order in which the expressions appear. Moreover, it is not efficient, for we have to store all the expressions and compare each result calculated with all the other stored expressions.

Also we want the representation to be "natural" (which today at least excludes the use of Roman numerals), and, at least in general, "compact" (which forbids the representation of integers in unary, so that 7 becomes 1111111). Fortunately there are many representations with these properties, and every Computer Algebra system has (at least!) one. Most of the systems (in particular REDUCE) always simplify (that is — put into canonical form): MACSYMA is an exception and only simplifies on demand.

For polynomials in one variable, these representations are, mathematically speaking, fairly obvious: every power of x appears only once (at most), and therefore the equality of polynomials comes down to the problem of equality of coefficients.

2.3.2 Dense and sparse representations

Now we have to distinguish between two types of representation: *dense* or *sparse*. Every Computer Algebra system has its own representation (or its own representations) for polynomials and the details only begin to be interesting when we have to tinker with the internal representation, but the distinction dense/sparse is important. A representation is called *sparse* if the zero terms are not explicitly represented. Conversely, a representation is called *dense* if all the terms (or at least those between the monomial of highest degree and the monomial of lowest degree) are represented — zero or non-zero. Normal mathematical representation is sparse: we write $3x^2 + 1$ instead of $3x^2 + 0x + 1$.

The most obvious computerised representation is the representation of a polynomial $a_0 + a_1 x + \cdots + a_n x^n$ by an array of its coefficients $[a_0, a_1, \ldots, a_n]$. All the a_i, zero or non-zero, are represented, therefore the representation is dense. In this representation, the addition of two polynomials of degree n involves $n + 1$ (that is $O(n)$) additions of coefficients, whereas multiplication by the naïve method involves $O(n^2)$ multiplications of coefficients (one can do it in $O(n \log n)$ [Aho *et al.*, 1974, Chapter 8], but no system uses this at present — see Probst and Alagar [1982] for an application and a discussion of the cost).

A less obvious representation, from the computing point of view, but closer to the mathematical representation, is a sparse representation: we store the exponent and the coefficient, that is the pair (i, a_i), for all the non-zero terms $a_i x^i$. Thus $3x^2 + 1$ can be represented as $((2, 3), (0, 1))$. This representation is fairly difficult in FORTRAN, it is true, but it is more natural in a language such as LISP (the language preferred by most workers in Computer Algebra). We stress the fact that the exponents must be ordered (normally in decreasing order), for otherwise $((2, 3), (0, 1))$ and $((0, 1), (2, 3))$ would be two different representations of the same polynomial. To prove that this representation is not very complicated, at least not in LISP, we give procedures for the addition and multiplication of polynomials with this representation. Readers who are not familiar with LISP can skip these definitions without losing the thread.

We shall use a representation in which the CAR of a polynomial is a term, whilst the CDR is another polynomial: the initial polynomial less the term defined by the CAR. A term is a CONS, where the CAR is the exponent and the CDR is the coefficient. Thus the LISP structure of the polynomial $3x^2 + 1$ is ((2 . 3) (0 . 1)), and the list NIL represents the polynomial 0. In this representation, we must note that the number 1 does not have the same representation as the polynomial 1 (that is ((0 . 1))), and that the polynomial 0 is represented differently from the other numerical polynomials.

```
(DE ADD-POLY (A B)
  (COND ((NULL A) B)
        ((NULL B) A)
        ((GREATERP (CAAR A) (CAAR B))
            (CONS (CAR A) (ADD-POLY (CDR A) B)))
        ((GREATERP (CAAR B) (CAAR A))
            (CONS (CAR B) (ADD-POLY A (CDR B))))
        ((ZEROP (PLUS (CDAR A) (CDAR B)))
            ; We must not construct a zero term
            (ADD-POLY (CDR A) (CDR B)))
        (T (CONS (CONS (CAAR A) (PLUS (CDAR A) (CDAR B)))
                 (ADD-POLY (CDR A) (CDR B))))))
(DE MULTIPLY-POLY (A B)
  (COND ((OR (NULL A) (NULL B)) NIL)
; If a = a0+a1 and b = b0+b1, then ab =
; a0b0 + a0b1 + a1b
        (T (CONS (CONS (PLUS (CAAR A) (CAAR B))
                       (TIMES (CDAR A) (CDAR B)))
                 (ADD-POLY (MULTIPLY-POLY (LIST (CAR A))
                                          (CDR B))
                           (MULTIPLY-POLY (CDR A) B))))))
```

If A has m terms and B has n terms, the calculating time (that is the number of LISP operations) for ADD-POLY is bounded by $O(m + n)$, and that for MULTIPLY-POLY by $O(m^2 n)$ ($(m(m + 3)/2 - 1)n$ to be exact). There are multiplication algorithms which are more efficient than this one: roughly speaking, we ought to sort the terms of the product so that they appear in decreasing order, and the use of ADD-POLY corresponds to a sorting algorithm by insertion. Of course, the use of a better sorting method (such as "quicksort") offers a more efficient multiplication algorithm, say $O(mn \log m)$ [Johnson, 1974]. But most systems use an algorithm similar to the procedure given above.

There is a technical difficulty with this procedure, discovered recently by Abbott *et al.* [1987]. We shall explain this difficulty in order to illustrate the problems which can arise in the translation of mathematical formulae into Computer Algebra systems. In MULTIPLY-POLY, we add $a_0 b_1$ to $a_1 b$. The order in which these two objects are calculated is important. Obviously this can change neither the results nor the time taken. But it can change the memory space used during the calculations. If a and b are dense of degree n, the order which first calculates $a_0 b_1$ should store all these intermediate results before the recursion finishes. Therefore the memory space needed is $O(n^2)$ words, for there are n results of length between 1 and n. The other order, $a_1 b$ calculated before $a_0 b_1$, is clearly more efficient, for the space used at any moment does not exceed $O(n)$. This is not a purely theoretical remark: Abbott *et al.* were able to factorise $x^{1155} - 1$ with REDUCE in 2 mega-bytes of memory, but they could not remultiply the factors without running out of memory.

In any case one might think that the algorithms which deal with sparse representation are less efficient that those which deal with dense representation: $m^2 n$ instead of mn for elementary multiplication or $mn \log m$ instead of $\max(m, n) \log \max(m, n)$ for advanced multiplication. It also looks as though there is a waste of memory, for the dense representation requires $n + 2$ words of memory to store a polynomial of degree n ($n + 1$ coefficients and n itself), whilst sparse representation requires at least $2n$, and even $4n$ with the natural method in LISP. In fact, for completely dense polynomials, this comparison is fair. But the majority of polynomials one finds are not dense. For example, a dense method requires a million multiplications to check

$$(x^{1000} + 1)(x^{1000} - 1) = x^{2000} - 1,$$

whereas a sparse method only requires four, for it is the same calculation as

$$(x + 1)(x - 1) = x^2 - 1.$$

When it is a question of polynomials in several variables, all realistic polynomials ought to be sparse — for example, a dense representation of $a^5 b^5 c^5 d^5 e^5$ contains 7776 terms.

Therefore, the calculating time, at least for addition and multiplication, is a function of the number of terms rather than of the degree. We are familiar with the rules bounding the degree of a result as a function of the degrees of the data, but the rules for the number of terms are different. Suppose that A and B are two polynomials, of degree n_A and n_B with m_A and m_B terms. Then the primitive operations of addition, subtraction, multiplication, division, calculation of the g.c.d. and substitution of a polynomial for the variable of another one satisfy the bounds below:

Operation	Degree of result	Number of terms in result
$A + B$	$\max(n_A, n_B)$	$m_A + m_B$
$A - B$	$\max(n_A, n_B)$	$m_A + m_B$
$A * B$	$n_A + n_B$	$m_A m_B$
A/B	$n_A - n_B$	$n_A - n_B + 1$
g.c.d.(A, B)	$\min(n_A, n_B)$	$\leq \min(n_A, n_B) + 1$
subst$(x = A, B)$	$n_A n_B$	$n_A n_B + 1$

The bound we give for the number of terms in the result of a division is a function of the degrees, and does not depend on the number of terms in the data. This may seem strange, but if we think of $(x^n - 1)/(x - 1) = x^{n-1} + x^{n-2} + \cdots + 1$, we see that two terms in each polynomial can produce a polynomial with an arbitrarily large number of terms.

For the g.c.d., the problem of limiting (non-trivially) the number of terms is unsolved*. For substitution, it is still possible for the result to be completely dense, even though the data are sparse. From this we deduce a very important observation: there may be a great difference between the time taken for calculating a problem in terms of x and the same problem in terms of $x - 1$ (or any other substitution).

To show what this remark means in practice, let us consider the matrix

$$A = \begin{pmatrix} 1 + 2x & x + x^4 & x + x^9 \\ x + x^4 & 1 + 2x^4 & x^4 + x^9 \\ x + x^9 & x^4 + x^9 & 1 + 2x^9 \end{pmatrix}.$$

We can calculate A^2 in .42 seconds (REDUCE on a micro-computer Motorola 68000), but if we rewrite it in terms of $y = x - 1$, this calculating time becomes 5.26 seconds — more than 10 times more. This problem comes up likewise in factorisation, as we shall see in the chapter "Advanced algorithms".

2.3.3 The g.c.d.

As we shall see later, calculations with rational fractions involve calculating the g.c.d. of the numerator and denominator of fractions. These calculations are less obvious than one might think, and to illustrate this remark let us consider the g.c.d. of the following two polynomials (this analysis is mostly taken from Brown [1971]) :

$$A(x) = x^8 + x^6 - 3x^4 - 3x^3 + 8x^2 + 2x - 5;$$
$$B(x) = 3x^6 + 5x^4 - 4x^2 - 9x + 21.$$

The first elimination gives $A - (\frac{x^2}{3} - \frac{2}{9})B$, that is

$$\frac{-5}{9}x^4 + \frac{1}{9}x^2 - \frac{1}{3},$$

---

* During the preparation of this translation, the authors were informed by B.M. Trager that A. Schinzel had shown that the polynomial

$$p = (1 + 2 * x - 2 * x^2 + 4 * x^3 + 4 * x^4)$$
$$* (1 + 2 * x^4 - 2 * x^8 + 4 * x^{12} - 10 * x^{16} + 28 * x^{20} - 84 * x^{24})$$

has the remarkable property that p^2 has *fewer* terms than p. Since p is square-free, this means that the g.c.d. of p^2 and $(p^2)'$ has more terms than either polynomial separately.

Coppersmith and Davenport have shown that, for any $n > 1$ and any $\epsilon > 0$, there is a polynomial p such that p^n has ϵ times as many terms as p.

and the subsequent eliminations give

$$\frac{-117}{25}x^2 - 9x + \frac{441}{25},$$

$$\frac{233150}{6591}x - \frac{102500}{2197},$$

and, finally,

$$\frac{1288744821}{543589225}.$$

It is obvious that these calculations on polynomials with rational coefficients require several g.c.d. calculations on integers, and that the integers in these calculations are not always small.

We can eliminate these g.c.d. calculations by working all the time with polynomials with integer coefficients, and this gives *polynomial remainder sequences* or *p.r.s.* Instead of dividing A by B in $\mathbf{Q}$, we can multiply A by a power (that is with exponent the difference of the degrees plus one) of the leading coefficient of B, so that this multiple of A can be divided by B over $\mathbf{Z}$. Thus we deduce the following sequence:

$$-15x^4 + 3x^2 - 9,$$

$$15795x^2 + 30375x - 59535,$$

$$1254542875143750x - 1654608338437500$$

and

$$12593338795500743100931151992187500.$$

These sequences are called *Euclidean sequences* (even though Euclid did not know about polynomials as algebraic objects). In general, the coefficients of such a sequence undergo an exponential increase in their length (as functions of the degrees of the given polynomials).

Obviously we can simplify by the common factors of these sequences (this algorithm is the algorithm of *primitive sequences*), but this involves the calculation of the g.c.d., which puts us back into the morass from which we have just emerged. Fortunately, it is possible to choose the coefficients by which we multiply A before dividing it by B etc., so that the increase of the coefficients is only linear. This idea, due independently to Brown and to Collins, is called *sequence of sub-resultant polynomials (sub-resultant p.r.s.)*. This algorithm is described by Brown [1971] and by Loos [1982], but it is important enough to be described here, even though we omit the proofs.

Suppose that the data are two polynomials F_1 and F_2 with integer coefficients. We determine their g.c.d. with the help of a sequence of polynomials $F_3, \ldots$. Let δ_i be the difference between the degrees of F_i and F_{i+1}, and f_i the leading coefficient of F_i. Then the remainder from the division of $f_i^{\delta_{i-1}+1} F_{i-1}$ by F_i is always a polynomial with integer coefficients. Let us call it $\beta_{i+1} F_{i+1}$, where β_{i+1} is to be determined. The algorithm called "Euclidean" corresponds to the choice of $\beta_{i+1} = 1$. If we put

$$\beta_3 = (-1)^{\delta_1+1},$$
$$\beta_i = -f_{i-2}\psi_i^{\delta_{i-2}}$$

where the ψ are defined by

$$\psi_3 = -1,$$
$$\psi_i = (-f_{i-2})^{\delta_{i-3}} \psi_{i-1}^{1-\delta_{i-3}},$$

then (by the *Sub-resultant Theorem* [Loos, 1982]) the F_i are always polynomials with integer coefficients, and the increase in length of the coefficients is only linear. For the same problem as before, we get the sequence:

$$F_3 = 15x^4 - 3x^2 + 9,$$
$$F_4 = 65x^2 + 125x - 245,$$
$$F_5 = 9326x - 12300,$$
$$F_6 = 260708,$$

and the growth is clearly less great than for the previous sequences.

This algorithm is the best method known for calculating the g.c.d., of all those based on Euclid's algorithm applied to polynomials with integer coefficients. In the chapter "Advanced algorithms" we shall see that if we go beyond these limits, it is possible to find better algorithms for this calculation.

2.4 POLYNOMIALS IN SEVERAL VARIABLES

There is a new difficulty as soon as we begin to calculate with polynomials in several variables: do we write $x+y$ or $y+x$? Mathematically, we are dealing with the same object, and therefore, if we want a canonical representation, we ought to write only one of these expressions. But which one? To be able to distinguish between them, we have to introduce a new idea — that of an order among the variables. We have already used, implicitly, the idea of an order among the powers of a variable, since we write $x^3 + x^2 + x^1 + x^0$ instead of $x^2 + x^0 + x^3 + x^1$, but this order seems "natural" (Norman [1982] has written a system which does not order the monomials according to this arrangement, but this gives rise to many unsolved problems). The idea of an order among the variables may seem artificial, but it is as necessary as the other. When we have decided on an order, such as "x is more important than y", we know that $x + y$ is the canonical representation, but that $y + x$ is not.

When it is a question of more complicated monomials (we recall that a monomial is a product of powers of variables, such as $x^2 y^3$ or $x^1 y^2 z^3$), there are various ways of extending this order. The most common ways are the following:

(a) We can decide that the main variable (the one which is in front of all the others in our order) is to determine the order as exactly as possible, and that we will only consider the powers of other variables if the powers of the first variable are equal. This system is called *lexicographic*, for it is the same system as the one used in a dictionary, where one looks first at the first letter of the two words, and it is only if they are the same that one looks at the second etc. In this system, the polynomial $(x + y)^2 + x + y + 1$ is written $x^2 + 2xy + x + y^2 + y + 1$ (if x is more "principal" than y).

(b) We can decide that the total degree of the monomial (that is the sum of the powers of the variables) is the most important thing, and that the terms of total degree 2 ought to appear before all the terms of total degree 1 etc. We use the previous method to distinguish between terms of the same total degree. This system is called *total degree*, or, more precisely, *total degree, then lexicographic*. In this system, the polynomial $(x + y)^2 + x + y + 1$ is written $x^2 + 2xy + y^2 + x + y + 1$ (if x is more "principal" than y).

(c) Instead of the lexicographic method, we can use the opposite. For a polynomial in one variable, this is the same as the increasing order of the powers, and this gives rise to difficulties (at least from the naïve point of view) in division and when calculating g.c.d.s. But this system has advantages when linked to the total degree method, as *total degree, then inverse lexicographic*. In this system, the polynomial $(x + y)^2 + x + y + 1$ is written $y^2 + 2xy + x^2 + y + x + 1$ (if x is more "principal" than y).

The reader might think that the systems (b) and (c) are equivalent, but with the order of the variables inverted. That is true for the case of two variables, but not for more than two. To show the difference, let us look at the expansion of $(x + y + z)^3$. First of all, in the order total degree then lexicographic (x before y before z), we get

$$x^3 + 3x^2 y + 3x^2 z + 3xy^2 + 6xyz + 3xz^2 + y^3 + 3y^2 z + 3yz^2 + z^3.$$

In the order total degree, then inverse lexicographic (z before y before x), we get

$$x^3 + 3x^2y + 3xy^2 + y^3 + 3x^2z + 6xyz + 3y^2z + 3xz^2 + 3yz^2 + z^3.$$

In this order, we have taken all the terms which do not involve z, before attacking the others, whereas the first order chose all the terms containing x^2 (even x^2z) before attacking the others.

In fact, deciding which of these systems is the best for a particular calculation is not always very obvious, and we do not have any good criteria for choosing (but see Buchberger [1981]). Most of the existing systems use a lexicographic order, but each system has its own particular methods. Lexicographic representation has an interesting consequence: as all the terms with x^n are grouped together (supposing that x is the main variable), we can therefore consider this polynomial as a polynomial in x, whose coefficients are polynomials in all the other variables. So the polynomial in case (a) can be written in the form $x^2 + x(2y+1) + (y^2+y+1)$. This form is called *recursive*, in contrast to the *distributed* form $x^2 + 2xy + x + y^2 + y + 1$. The recursive form is used by most systems.

This discussion provides us with some indications about the behaviour of systems. For example, in MACSYMA, the function INPART lets us choose one part of an expression, but it operates on the internal form, which is more or less inverse lexicographic recursive. Similarly, REDUCE has a function COEFF, which gives the coefficients of an expression, seen as a polynomial in one named variable. As the internal representation of REDUCE is lexicographic recursive, it is obvious that this function is much quicker when the named variable is the principal one. For example, if we take $(w+x+y+z)^6$, the time of COEFF varies between .50 seconds (w named) and .84 seconds (z named). In general, the calculation time (and the work memory) may vary greatly according to the order of the variables [Pearce and Hicks, 1981, 1982, 1983], but the reasons are not always very obvious. We do not know of any good general rules for choosing the order, for it is largely determined by the form desired for printing the results. We must also point out that changing the order is expensive, since all the results which have already been calculated have to be re-expressed in the new order.

These questions of order are not just internal details which concern only the programmers: they can influence the results obtained. For example, suppose we want to divide $2x - y$ by $x + y$. If x is the main variable, then the quotient is 2, and the remainder is $-3y$. But if y is the main variable, then the quotient is -1, and the remainder is $3x$.

2.5 REPRESENTATIONS OF RATIONAL FUNCTIONS

Most calculations use not only polynomials, but also fractions of them, that is rational functions. The same remarks which applied in the polynomial case apply here: the calculation

$$\frac{1}{\sin x} + \frac{1}{\cos x} = \frac{\cos x + \sin x}{\cos x \sin x}$$

is a rational calculation — in fact it is the same calculation as

$$\frac{1}{b} + \frac{1}{a} = \frac{a+b}{ab}.$$

If we represent a rational function by a polynomial (the *numerator*) divided by another polynomial (the *denominator*), we have of course a normal representation, for the function represents zero if and only if its numerator is zero. With REDUCE, it is possible not to use this representation (OFF MCD), but we do not advise this, because we no longer know whether a formula represents zero or not, and this causes many problems with Gaussian elimination, for example in

$$\begin{pmatrix} 1 & 1/(x-1) \\ 1/(x+1) & 1/(x^2-1) \end{pmatrix}.$$

If we want a canonical representation, we have to do more than express in the form of the quotient of two polynomials. For example, the formulae $(x-1)/(x+1)$ and $(x^2 - 2x + 1)/(x^2 - 1)$ represent the same element of $\mathbf{Q}(x)$, but they are two different formulae.

Here we must say a few words about the difference between the elements of $\mathbf{Q}(x)$ and the functions of $\mathbf{Q}$ into itself which they represent. Seen as functions of $\mathbf{Q}$ into itself, $f(x) = (x-1)/(x+1)$ and $g(x) = (x^2 - 2x + 1)/(x^2 - 1)$ are two different functions, for the first is defined at the value $x = 1$ (where it takes the value 0), whereas $g(x)$ is not defined, since it becomes 0/0. But this singularity is not "intrinsic", because the function defined by

$$g_1(x) = \begin{cases} \dfrac{x^2 - 2x + 1}{x^2 - 1} & (x \neq 1) \\ 0 & (x = 1) \end{cases}$$

is continuous, differentiable etc. at $x = 1$. In general, it is not very hard to check that, if $f_1 = p_1/q_1$ is a simplification of $f = p/q$ (where p, q, p_1 and q_1 are polynomials), then at each value x_0 of x where f is defined, $f_1(x_0)$ is defined and equal to $f(x_0)$, and, moreover, if $f(x_1)$ is not defined, but $f_1(x_1)$ is defined, then the function

$$g(x) = \begin{cases} f(x) & (x \neq x_1) \\ f_1(x) & (x = x_1) \end{cases}$$

is continuous etc. at $x = x_1$. Thus this kind of simplification of a formula only changes the function by eliminating such singularities*.

Seen algebraically, i.e. as elements of $\mathbf{Q}(x)$, the expressions $(x-1)/(x+1)$ and $(x^2 - 2x + 1)/(x^2 - 1)$ are equal, since the difference between them is $0/(x^2 - 1) = 0$. Thus there is a slight distinction between the elements of $\mathbf{Q}(x)$ and the functions of $\mathbf{Q}$ into itself, but, with the usual abuse of language, we call the elements of $\mathbf{Q}(x)$ rational functions. (Bourbaki : "... the abuses of language without which any mathematical text threatens to become pedantic and even unreadable").

After this short digression, we return to the problem of a canonical representation for rational functions, that is to say, the elements of $\mathbf{Q}(x)$. We have established that $(x-1)/(x+1)$ and $(x^2 - 2x + 1)/(x^2 - 1)$ are two different representations of the same function. The definition of "canonical" implies the fact that at least one of them is not canonical. The natural choice is to say that there must not be any divisor common to the numerator and the denominator. This implies that $(x^2 - 2x + 1)/(x^2 - 1)$ is not canonical, for there is a g.c.d. which is $(x - 1)$. If we remove this g.c.d., we come back to $(x-1)/(x+1)$. In general, we find a representation with the degree of the numerator as small as possible (and the same for the denominator). If there were only one such representation, it would be a good choice for a canonical representation. Unfortunately, the condition that the degree of the numerator is to be minimised does not give us uniqueness, as the following examples show:

$$\frac{-2x+1}{2x+1} = \frac{2x-1}{-2x-1} = \frac{4x-2}{-4x-2} = \frac{-x+1/2}{x+1/2} = \frac{x-1/2}{-x-1/2}.$$

To resolve these ambiguities, most existing systems take into account the following rules (for rational functions with coefficients in $\mathbf{Q}$):
(1) no rational coefficient in the expression (which eliminates the last two expressions);
(2) no integer may divide both the numerator and the denominator of the expression (which eliminates $(4x - 2)/(-4x - 2)$);
(3) the leading coefficient of the denominator of the expression must be positive (which eliminates the second expression).
There are several other possibilities, but these rules are the ones most used and are sufficient to give a canonical form. As we have already said in the case of the rational numbers, the fact that one *can* calculate with rational fractions does not mean that one *should* calculate with them. If one can

---

* The reader who is familiar with the theory of denotational semantics can express these ideas differently by noting that f_1 is higher (in the lattice of partial functions of $\mathbf{Q}$ into itself) than the function f.

find an algorithm which avoids these calculations, it will in general be more efficient. We shall see later that Bareiss' algorithm is a variation of Gaussian elimination which does not need the use of fractions.

2.6 REPRESENTATIONS OF ALGEBRAIC FUNCTIONS

We understand by *algebraic* a solution of a polynomial equation. $\sqrt{2}$ is an algebraic number, since it is both a number and a solution of the equation $\alpha^2 - 2 = 0$. This equation has two roots, but we shall not differentiate between them in this section. In the next chapter we shall see how to differentiate between the different real values which satisfy the same equation, but this is unnecessary for many applications. $\sqrt[3]{x^2 - 1}$ is an algebraic function, because it is both a function and a solution of the equation $\beta^3 - x^2 + 1 = 0$. Almost all this section applies to both functions and numbers: in general we shall speak of algebraic functions, even though most of the examples are algebraic numbers. Every radical is an algebraic expression, but the opposite is false, as the great mathematician Abel proved. More precisely, the algebraic number γ which is a solution of $\gamma^5 + \gamma + 1 = 0$ cannot be expressed in radicals.

One can therefore distinguish three classes of algebraic expressions:
(1) the simple radicals, such as $\sqrt{2}$ or $\sqrt[3]{x^2 - 1}$;
(2) the simple or nested radicals, which include also expressions such as $\sqrt{1 + \sqrt{2}}$ or $\sqrt[3]{\sqrt{2} + \sqrt[3]{x}}$;
(3) the general algebraic expressions, which also include expressions such as the algebraic number γ defined by $\gamma^5 + \gamma + 1 = 0$.
Each class is contained in the subsequent ones.

2.6.1 The simple radicals

For the first class, the representation is fairly obvious: we consider each radical as a variable appearing in a polynomial or rational expression. Obviously, if α is an n-th root, we only use the powers $0, \ldots, n - 1$ of α, and we replace the higher powers by lower powers.

This representation is not canonical for two different reasons. In the first place there is a problem with rational fractions which contain algebraic expressions. Let us look for example at $1/(\sqrt{2} - 1)$ and $\sqrt{2} + 1$. These two expressions are not equal, but they represent the same number, for their difference is zero:

$$\frac{1}{\sqrt{2} - 1} - (\sqrt{2} + 1) = \frac{1}{\sqrt{2} - 1} - \frac{(\sqrt{2} + 1)(\sqrt{2} - 1)}{\sqrt{2} - 1}$$

$$= \frac{1 - (\sqrt{2})^2 + 1}{\sqrt{2} - 1} = \frac{0}{\sqrt{2} - 1} = 0.$$

One solution is to insist that the roots appear only in the numerator: this forbids the representation $1/(\sqrt{2}-1)$. For every root α, we can achieve this by multiplying the numerator n and the denominator d of the expression by a polynomial d_1, such that the algebraic quantity α does not appear in the product dd_1. In this case,

$$\frac{n}{d} = \frac{nd_1}{dd_1}$$

and α is removed from the denominator. We can calculate this polynomial d_1 by applying the extended Euclidean algorithm (see the appendix "Algebraic background") to the pair d and p, where p is the polynomial defining α. This application produces two polynomials d_1 and d_2 such that $dd_1 + pd_2 = c$, where c does not depend on α, and then $dd_1 = c$.

With this restriction, we can be sure that we shall have a canonical system, if the radicals form an independent system. But this representation is not always very efficient: for example $\dfrac{1}{\sqrt{2}+\sqrt{3}+\sqrt{5}+\sqrt{7}}$ becomes

$$\frac{\begin{aligned}22\sqrt{3}\sqrt{5}\sqrt{7} - 34\sqrt{2}\sqrt{5}\sqrt{7} - 50\sqrt{2}\sqrt{3}\sqrt{7} + 135\sqrt{7}+ \\ 62\sqrt{2}\sqrt{3}\sqrt{5} - 133\sqrt{5} - 145\sqrt{3} + 185\sqrt{2}\end{aligned}}{215}.$$

Because of this growth, we often stop at the normal representation given by rational fractions in the powers of α less than n.

But even with this restriction to polynomials, there is another problem: that of the interdependence of the radicals. A very simple example of this kind of problem is that we must not construct $\sqrt{1}$, which is equal to 1. Similarly, we can calculate with $\sqrt{2}$, or with $\sqrt{8}$, but not with both of them, for $\sqrt{8} = 2\sqrt{2}$, and this representation is not canonical and not even normal. Similarly, among the radicals $\sqrt{2}$, $\sqrt{3}$ and $\sqrt{6}$, we can work with any two out of the three: if they are all present we get the relation $\sqrt{2}\sqrt{3} = \sqrt{6}$. All these examples are "obvious", and we might suppose that it is enough to make sure that all the numbers (or polynomials or rational functions) which are in the radicals are relatively prime, and that the radicals cannot be simplified. Unfortunately, things are a little more complicated. Let us consider $\alpha = \sqrt[4]{-4}$. One might suppose that there is no possible simplification, but in fact $\alpha^2 = 2\alpha - 2$. This is a consequence of the factorisation $x^4 + 4 = (x^2 - 2x + 2)(x^2 + 2x + 2)$. Following Capelli [1901], we can prove that this example (and variations on it such as $\sqrt[4]{-4.3^4}$) is the only possible non-trivial simplification. Najid-Zejli [1984, 1985] has studied these questions, and has given an algorithm to decide if there is a relation between non-nested radicals.

2.6.2 Nested radicals

The problems are more difficult for the second class, i.e. the nested radicals. The first problem, that of the equivalence between rational fractions which contain radicals, has the same "solution" as before, and the same defects. The other problem, that of relations between the radicals, is still not solved (in a satisfactory manner). There is a general solution, but it is the same solution as for general algebraic expressions, and we shall consider it later. We cannot prove that this problem is difficult, but there are many surprising examples. We cite the identities

$$\sqrt{9 + 4\sqrt{2}} = 1 + 2\sqrt{2} \tag{1}$$

$$\sqrt{5 + 2\sqrt{6}} + \sqrt{5 - 2\sqrt{6}} = 2\sqrt{3} \tag{2}$$

$$\sqrt{x + \sqrt{x^2 - 1}} = \sqrt{\frac{x+1}{2}} + \sqrt{\frac{x-1}{2}} \tag{3}$$

$$\sqrt{16 - 2\sqrt{29} + 2\sqrt{55 - 10\sqrt{29}}} = \sqrt{22 + 2\sqrt{5}} - \sqrt{11 + 2\sqrt{29}} + \sqrt{5} \tag{4}$$

$$\sqrt[3]{\sqrt[5]{32/5} - \sqrt[5]{27/5}} = \sqrt[5]{\frac{1}{25}} + \sqrt[5]{\frac{3}{25}} - \sqrt[5]{\frac{9}{25}}$$

$$= \sqrt[5]{\frac{1}{25}} \left(1 + \sqrt[5]{3} - \sqrt[5]{3}^2\right) \tag{5}$$

$$\sqrt{(112 + 70\sqrt{2}) + (46 + 34\sqrt{2})\sqrt{5}} = (5 + 4\sqrt{2}) + (3 + \sqrt{2})\sqrt{5} \tag{6}$$

where we owe (1) to Davenport [1981], (2) and (3) to Zippel [1985], (4) to Shanks [1974], (5) to Ramanujan [1927] and (6) to Borodin *et al.* [1985]. Recently, Borodin *et al.* [1985] and Zippel [1985] have been studying this problem, and they have found algorithms which can solve several nested radicals, either by writing them in a non-nested way, or by proving that there is no simplification of this kind. But these algorithms are limited to a few special cases, such as square roots with two levels of nesting. The general case remains, as we said earlier, unresolved (in a satisfactory way).

2.6.3 General algebraic functions

Now we look at algebraic functions which are defined as being the roots of some polynomial. It is possible that there exists a representation in terms of radicals, but, as we have already stated, it is also possible that there is not one. So, let α be an algebraic function (or number) defined as the root of the polynomial $p(\alpha)$. If the polynomial p is not irreducible, even simple calculations confront us with many problems. For example, take $p(\alpha) = \alpha^2 - 3\alpha + 2$. Thus α is an algebraic number of degree two. $(\alpha - 1)$ and $(\alpha - 2)$ are two numbers, *a priori* non-zero. But their product is $p(\alpha)$, which reduces to zero. In fact, this is a generalisation of the problem we came up against with simple radicals, where $\sqrt{1}$ was not a legitimate radical. In this case, seen from this new angle, the problem is that the polynomial defining $\sqrt{1}$, i.e. $\alpha^2 - 1$, is not irreducible and has a factor $\alpha - 1$. In the same way, our new polynomial has factors $\alpha - 1$ and $\alpha - 2$, and therefore one of the expressions $(\alpha - 1)$ and $(\alpha - 2)$ is zero.

Even if the factors are not linear, any calculation with the roots of a reducible polynomial may result in an impasse, where two expressions, apparently non-zero, give zero when multiplied. The other difficulty which may arise is the impossibility of dividing by a non-zero expression, for it may have a non-trivial g.c.d. with the polynomial defining α. Therefore we require, generally speaking, that the polynomials defining the algebraic numbers and functions be irreducible, or else we admit that we cannot guarantee the results if the polynomials are not irreducible. (But note that each "difficulty" we have mentioned earlier gives rise to a factorisation of these polynomials, and so it is possible to use reducible polynomials to create algebraic extensions. See Della Dora *et al.* [1985] and Dicrescenzo and Duval [1985] for the details.)

All the examples of the preceding sub-sections, where the radicals were simpler than had been thought, have in common the fact that the polynomials were not irreducible. $\sqrt{1}$ is only an integer, because the polynomial which seems to define it, that is $\alpha^2 - 1$, factorises into $(\alpha - 1)(\alpha + 1)$. Similarly, $\sqrt[4]{-4}$ is not legitimate, for its polynomial, $\alpha^4 + 4$, factorises into $(\alpha^2 - 2\alpha + 2)(\alpha^2 + 2\alpha + 2)$.

The same is true for nested radicals. We consider equation (1) of the previous sub-section, where the polynomial defining $\sqrt{9 + 4\sqrt{2}}$ is $\alpha^2 - (9 + 4\sqrt{2})$. This polynomial factorises into $(\alpha - (1 + 2\sqrt{2}))(\alpha + (1 + 2\sqrt{2}))$, and therefore the nested radical has a simpler form. If we want to treat only polynomials with integer coefficients, we can say that α is a root of $\alpha^4 - 18\alpha^2 + 49$ (which is the *norm** of the polynomial already given). This polynomial also factorises into $(\alpha^2 - 2\alpha - 7)(\alpha^2 + 2\alpha - 7)$, and the roots of the first factor are $1 \pm 2\sqrt{2}$.

Thus, a definition such as "α a root of $p(\alpha)$" is legitimate if and only if p is irreducible. Here "legitimate" means that the use of polynomials in α (of degree less than that of p) gives a canonical representation, and that the use of such rational functions (which do not contain this root in the denominator) also gives a canonical representation.

---

* These *norms* can be calculated using resultants — see the appendix "Algebraic background".

When several roots appear, obviously all the polynomials defining them must be irreducible. But we need more than that, in the sense that they must be irreducible, not only *separately*, but also *together*. Consider, for example, α a root of $\alpha^5 - \alpha - 1$, and β a root of $\beta^5 + 5\beta^4 + 10\beta^3 + 10\beta^2 + 4\beta - 1$. These two polynomials are irreducible, when viewed as polynomials with integer coefficients. But if we take the polynomial defining β as a polynomial whose coefficients may depend on α, the situation is very different. In fact this polynomial factorises into

$$(\beta - \alpha + 1)$$
$$(\beta^4 + \beta^3(\alpha+4) + \beta^2(\alpha^2+3\alpha+6) + \beta(\alpha^3+2\alpha^2+3\alpha+4) + \alpha^4 + \alpha^3 + \alpha^2 + \alpha).$$

This factorisation should not surprise us, because β is only $\alpha - 1$, as the linear factor above proves. (The other factor corresponds to the fact that $\alpha^5 - \alpha - 1$ has five roots, and β can be expressed in terms of a different root from the root chosen as α.)

Thus, to verify that a system of roots α_i of polynomials p_i is legitimate, we have to factorise p_1 as a polynomial with integer coefficients, then p_2 as a polynomial with coefficients in $\mathbf{Z}[\alpha_1]$, then p_3 as a polynomial with coefficients in $\mathbf{Z}[\alpha_1, \alpha_2]$, then $\dots$. If all the p_i are polynomials with integer coefficients, the order among the p_i does not change the result. If a p_i depends on an α_j, p_j has to be factorised before p_i. Note that, in fact, this process can easily be very expensive, for factorisations over algebraic fields are very hard to carry out.

2.6.4 Primitive elements

Instead of studying several algebraic numbers (or functions), such as $\sqrt{2}$ and $\sqrt{3}$, we can always* go back to the case of a single algebraic number (or of a single algebraic function), in terms of which all the others can be expressed. This quantity is called a *primitive element* for the field generated by the given quantities, or, more simply, a primitive element for the quantities themselves.

For example, the number α defined as a root of the polynomial $\alpha^4 - 10\alpha^2 + 1$ is $\sqrt{2} + \sqrt{3}$, and in terms of α, $\sqrt{2} = (\alpha^3 - 9\alpha)/2$ and $\sqrt{3} = (11\alpha - \alpha^3)/2$. α is therefore a primitive element for the field $\mathbf{Q}[\sqrt{2}, \sqrt{3}]$. These primitive elements can be calculated from the polynomials which define the given quantities, by using the resultant (see the appendix "Algebraic

---

* Given that we are working in a field with characteristic zero, that is an extension of the integers. This theorem does not hold if we are working in an extension of the integers modulo p, but we are not interested in this at present.

background"). For example, α is defined by the resultant of $x^2 - 3$ and $(x - y)^2 - 2$. We can already see that the relation between α and $\sqrt{2}$ and $\sqrt{3}$ is not obvious.

The primitive elements are often very complicated. Najid-Zejli [1985] noted that the primitive element corresponding to two roots α and β of the polynomial $x^4 + 2x^3 + 5$ is (at least if we calculate with the well-known algorithms [Trager, 1976]) a root of

$$\gamma^{12} + 18\gamma^{11} + 132\gamma^{10} + 504\gamma^9$$
$$+ 991\gamma^8 + 372\gamma^7 - 3028\gamma^6 - 6720\gamma^5$$
$$+ 11435\gamma^4 + 91650\gamma^3 + 185400\gamma^2 + 194400\gamma + 164525.$$

This polynomial is itself discouraging enough, but, in addition, the expressions for α and β in terms of γ require numbers with fourteen digits. When we are dealing with a primitive element for three of the roots (which is at the same time a primitive element for all the roots), the corresponding polynomial has coefficients of more than 200 digits.

We can conclude that, although primitive elements are fairly useful theoretically, they are too difficult to calculate and to use in practice.

2.7 REPRESENTATIONS OF TRANSCENDENTALS

Transcendental functions group together several classes of functions, each with its own special rules. In general, a function such as $\sin x$ is represented by a structure such as (SIN X) in LISP, or a "record" in PASCAL. Numbers such as $\sin 1$ or $\sin \pi$ can also be represented in this way. Thus, this structure is regarded as a variable and can appear in polynomials or rational functions. In REDUCE, for example, such a structure is called a "kernel". We have already seen that polynomial or rational calculations can apply to variables of this kind. The great problem then is the simplification of these variables, and their relation to one another. We know a lot of rules of simplification, such as

$$\sin(x + y) = \sin x \cos y + \cos x \sin y \tag{1}$$

$$\sin x \cos y = \frac{\sin(x + y) + \sin(x - y)}{2} \tag{2}$$

$$\log(xy) = \log x + \log y \tag{3}$$

$$\log \exp x = x \tag{4}$$

$$\exp(x + y) = \exp x \exp y \tag{5}$$

$$\sin \pi = 0. \tag{6}$$

Most Computer Algebra systems let the user define rules of this kind, which the system will take into account. For example, in REDUCE we can express the rules (1)–(6) by:

```
FOR ALL X,Y LET SIN(X+Y) = SIN(X)*COS(Y)+COS(X)*SIN(Y);
FOR ALL X,Y LET SIN(X)*COS(Y) = (SIN(X+Y)+SIN(X-Y))/2;
FOR ALL X,Y LET LOG(X*Y) = LOG(X) + LOG(Y);
FOR ALL X LET LOG(EXP(X)) = X;
FOR ALL X,Y LET EXP(X+Y) = EXP(X)*EXP(Y);
LET SIN(PI)=0;
```

There is a difference between the last rule and the others: the last one applies to a special number, whilst the others apply to any possible value of X or Y, which is expressed by the preamble FOR ALL.

Nevertheless, there are still some pitfalls in this area of rules (or, more exactly, rewrite rules). Firstly, we see that rule (1) implies that $\sin 2x = 2 \sin x \cos x$, whereas its expression in REDUCE does not have the same effect, for REDUCE does not see that $2x = x + x$. Therefore a better translation of the first rule would be

```
FOR ALL X,Y LET SIN(X+Y) = SIN(X)*COS(Y)+COS(X)*SIN(Y);
FOR ALL X LET SIN(2*X) = 2*SIN(X)*COS(X);
```

but even this is not enough (because of $\sin 3x$ etc.), and we have to add a rule such as

```
FOR ALL X,N SUCH THAT NUMBERP N AND N>1 LET
   SIN(N*X) = SIN((N-1)*X)*COS(X) + COS((N-1)*X)*SIN(X);
```

with corresponding rules for cos.

Secondly, rules (1) and (2) are mutual inverses. If we ask REDUCE to apply both, it loops* when we enter $\sin(a + b)$, for rule (1) rewrites it in the form $\sin a \cos b + \cos a \sin b$, and the first term of this is rewritten by rule (2) in the form $\frac{1}{2}(\sin(a + b) + \sin(a - b))$, which contains the original term, to which rule (1) can be applied again.

Thirdly, this simplification by rewrite rules can be very expensive. Every rewrite needs a resimplification (from the polynomial point of view) of the expression we are trying to simplify. Moreover, the simplification we have just done may give rise to other simplifications.

Fourthly, we are not certain that we have given all the necessary rules. Because of the interactive nature of Computer Algebra systems, this is not always a serious problem, but often we want all the trigonometric functions to be linearised, or all the logarithms to be independent etc.

All these problems can be linked to the fact that we are using a general method, that is the method of rewrite rules, to solve a problem which is clearly less general, such as the simplification of logarithmic or trigonomet-

---

* In principle. In fact, the present version of REDUCE notices that it has applied more rules than a system limit allows, stops and gives an error message.

ric functions. If we have a function of which we know nothing but some rules, the approach by rewrite rules is the only possible one.

There is a difference between knowing some possible simplifications (which may be rewrite rules), and knowing not only these simplifications, but also that they are the only possible ones. For example, we are familiar with the rules:

$$\log(fg) = \log f + \log g; \quad \exp(f + g) = \exp f * \exp g;$$
$$\exp \log f = \log \exp f = f; \tag{1}$$

but are they the only possible simplifications?

There are theorems which can describe precisely the possible simplifications. The first, and the one which covers the most important case, is Risch's structure theorem [Risch, 1979; Rosenlicht, 1976], which says, in effect, that the rules (1) are the only possible simplifications for functions generated by the operators exp and log. Although we are referring to fairly recent literature, the theorem (or, more exactly, the underlying principles) has been known since Liouville, but it is only Computer Algebra which requires such theorems to be explicit.

Structure Theorem. *Let K be a field of constants and $\theta_1, \ldots, \theta_n$ algebraic, exponential (when we write $\theta_i = u_i = \exp v_i$) or logarithmic (when we write $\theta_i = v_i = \log u_i$) functions, with each θ_i defined over $K(x, \theta_1, \ldots, \theta_{i-1})$, and with $K(x, \theta_1, \ldots, \theta_n)$ having K as the field of constants.*

(a) *In these conditions, a θ_i which is an exponential ($\theta_i' = v_i' \theta_i$) is transcendental over $K(x, \theta_1, \ldots, \theta_{i-1})$ if, and only if, v_i cannot be expressed as $c + \sum_{j=1}^{i-1} n_j v_j$, where c belongs to K and the n_i are rational numbers.*

(b) *Similarly, a θ_i which is a logarithm ($\theta_i' = u_i'/u_i$) is transcendental over $K(x, \theta_1, \ldots, \theta_{i-1})$ if, and only if, no power u_i^n of u_i can be expressed as $c \prod_{j=1}^{i-1} u_j^{n_j}$, where c belongs to K and n and the n_j are integers (with $n \neq 0$).*

This theorem may seem fairly complicated, but it can be re-expressed informally in a much simpler form:

(a) An exponential function is independent of the exponentials and logarithms which have already been introduced if, and only if, its argument cannot be expressed as a linear combination (with rational coefficients) of the logarithms and arguments of the exponentials which we have already introduced. Such a combination means that the new exponential is a product of powers of the exponentials and of the arguments of the logarithms already introduced.

(b) A logarithmic function is independent of the exponentials and logarithms already introduced if, and only if, its argument cannot be expressed as a product (with rational exponents) of the exponentials and of the arguments of the logarithms already introduced. Such a product means that the new logarithm is a linear combination (with rational coefficients) of the logarithms and arguments of the exponentials already introduced.

This theorem only applies to functions. The position for numbers defined by exponentials and logarithms is much less clear. It is conjectured that this theorem continues to hold, but we have no idea how to prove that. We do not even know whether e ($= \exp(1)$) and π ($= (1/i) \log(-1)$) are independent or not.

2.8 REPRESENTATIONS OF MATRICES

There are two styles of matrix calculations, which can be called *implicit* and *explicit*. An example of implicit calculation is provided by the mathematical expression "Let A and B be two square matrices of the same dimension". Here we have not defined exactly the dimension of the matrices and we have said nothing of their elements. On the other hand, in explicit calculation, we define exactly all the elements of the matrix, which may be, not only numbers, but also polynomials, rational functions, or any symbolic objects.

In fact, in implicit calculation A and B are variables. But the polynomial or rational calculation we have already seen does not apply to such variables, for they are *non-commutative variables*. For example AB may be different from BA. Several Computer Algebra systems let the user work with such variables. In MACSYMA, for example, we saw in the examples in Chapter 1 that there are two different operators for multiplication: $A*B$ for commutative multiplication and $A.B$ for non-commutative multiplication. Thus $A*B - B*A$ becomes 0, but $A.B - B.A$ remains unchanged. In REDUCE (see also the description of REDUCE's operators in the Annex), the possibilities are similar, but the manner of expressing them is different. From the moment we declare `NONCOM M`, the kernels which begin with `M` (such as `M(1)` or `M(A,B)`) will be non-commutative kernels, and will not commute with other non-commutative kernels, but will commute with ordinary (commutative) kernels. We can use rewrite rules to say that some non-commutative objects satisfy certain constraints.

In the remainder of this section, we shall deal with explicit matrix calculation. Here, as with polynomials, there is the distinction dense/sparse.

2.8.1 Dense matrices

The obvious way of representing an explicit matrix is an array of the elements of the matrix (if the implementation language does not have arrays, which is the case with several dialects of LISP, we can use vectors of vectors, or even lists of lists, but a representation by lists is less efficient). If the signs <...> signify a vector, then the matrix

$$\begin{pmatrix} a & b & c \\ d & e & f \\ g & h & i \end{pmatrix}$$

will be represented by < < a b c > < d e f > < g h i > > . For dense matrices, this method works quite well, and most Computer Algebra systems use it. The algorithms for the addition and multiplication of these matrices are the same as for numerical matrices, and imply that one can add two matrices of size n in $O(n^2)$ operations, and multiply them in $O(n^3)$ operations. As with numerical matrices, there are "non-obvious" algorithms for multiplication, which are, asymptotically, more efficient than usual algorithms, such as those of Strassen [1969] (which gives us an algorithm of complexity $O(n^{\log_2 7})$), Winograd [1968] and Coppersmith and Winograd [1982]. For numerical matrices, these methods are only quicker from $n > 20$ on, and are at most 18% quicker when $n = 100$ [Brent, 1970]. It is probable that the same conclusions hold to a large extent for the matrices of Computer Algebra. We do not know of any system which uses these "fast" methods.

When it comes to inversion, and problems associated with it such as the solution of linear systems and finding the determinant, the algorithms of numerical calculation are not easy to apply, for the difficulties which arise are very different in Computer Algebra and numerical calculation. In the first place, Computer Algebra does not have any problem of numerical stability, and therefore every non-zero element is a good pivot for Gaussian elimination. In this respect, Computer Algebra is simpler than numerical calculation.

On the other hand, there is a big problem with the growth of the data, whether they be intermediate or final. For example, if we take the generic matrix of size three, that is

$$\begin{pmatrix} a & b & c \\ d & e & f \\ g & h & i \end{pmatrix},$$

its determinant is

$$aei - afh - bdi + bfg + cdh - ceg,$$

and its inverse is

$$\frac{1}{aei - afh - bdi + bfg + cdh - ceg} \begin{pmatrix} ei - fh & -bi + ch & bf - ce \\ -di + fg & ai - cg & -af + cd \\ dh - eg & -ah + bg & ae - bd \end{pmatrix}.$$

For the generic matrix of size four, the determinant is

$$
\begin{aligned}
&afkp - aflo - agjp + agln + ahjo - ahkn - bekp + belo \\
&+bgip - bglm - bhio + bhkm + cejp - celn - cfip + cflm \\
&+chin - chjm - dejo + dekn + dfio - dfkm - dgin + dgjm,
\end{aligned}
$$

and the inverse is too large to be printed here. Therefore, in general, the data swell enormously if one inverts generic matrices, and the same is true for determinants of generic matrices, or for solutions of generic linear systems. Such results may appear as intermediate data in a calculation, the final result of which is small, for example in the calculation of the determinant, equal to 0, of a matrix of type $\begin{pmatrix} M & M \\ M & M \end{pmatrix}$, where M is a generic matrix.

The other big problem, especially for the calculation of determinants, is that of division. According to Cramer's rule, the determinant of a matrix is a sum (possibly with negations) of products of the elements of the matrix. Thus, if the elements belong to a ring (such as the integers or the polynomials), the determinant also belongs to it. But the elimination method requires several divisions. These divisions may not be possible (for example one cannot divide by 5 or 2 in the ring of integers modulo 10, but nevertheless the matrix $\begin{pmatrix} 5 & 2 \\ 2 & 5 \end{pmatrix}$ has a well defined determinant, that is, 1). Even if the divisions are possible, they imply that we have to calculate with fractions, which is very expensive because of the necessary g.c.d.s (which are often non-trivial).

Bareiss [1968] has described a cunning variation of Gaussian elimination, in which each division in the ring has to give a result in the ring, and not a fraction. This method (described in the next sub-section) is very often used in Computer Algebra, if the ring of the elements allows division (more precisely, if the ring is integral). Otherwise, there is a method found by Sasaki and Murao [1981, 1982] where one adds several new variables to the ring, and where one keeps only a few terms in these variables.

Cramer's method, which writes the determinant of a matrix of size n as the sum of $n!$ products of n elements of the matrix, is very inefficient numerically: the number of operations is $O(n(n!))$, instead of $O(n^3)$ for Gaussian elimination. But, in Computer Algebra, the cost of an operation depends on the size of the data involved. In this expansion, each intermediate calculation is done on data which are (at least if there is no cancellation) smaller than the result. It seems that in the case of a matrix of polynomials in several variables (these polynomials have to be sparse, otherwise the cost would be enormous), this method is clearly much faster than any method based on Gaussian elimination. For the case of a matrix of integers or of polynomials in one variable, Bareiss' method seems to be the most efficient.

2.8.2 Bareiss' algorithm

In fact, Bareiss has produced a whole family of methods for elimination without fractions, that is, where all the divisions needed are exact. These methods answer the problem stated in the section "Representations of fractions", the problem of finding an algorithm which does not require calculations with fractions. The simplest method, called "one step", which has in fact been known since Jordan, is based on a generalisation of Sylvester's identity. Let $a_{i,j}^{(k)}$ be the determinant

$$
\begin{vmatrix}
a_{1,1} & a_{1,2} & \cdots & a_{1,k} & a_{1,j} \\
a_{2,1} & a_{2,2} & \cdots & a_{2,k} & a_{2,j} \\
\cdots & \cdots & \cdots & \cdots & \cdots \\
a_{k,1} & a_{k,2} & \cdots & a_{k,k} & a_{k,j} \\
a_{i,1} & a_{i,2} & \cdots & a_{i,k} & a_{i,j}
\end{vmatrix}.
$$

In particular, the determinant of the matrix of size n (whose elements are $(a_{i,j})$) is $a_{n,n}^{(n-1)}$. The basic identity is the following:

$$
a_{i,j}^{(k)} = \frac{1}{a_{k-1,k-1}^{(k-2)}}
\begin{vmatrix}
a_{k,k}^{(k-1)} & a_{k,j}^{(k-1)} \\
a_{i,k}^{(k-1)} & a_{i,j}^{(k-1)}
\end{vmatrix}.
$$

In other words, after an elimination, we can be certain of being able to divide by the pivot of the preceding elimination.

To demonstrate Bareiss' method, let us consider a generic matrix of size three, that is:

$$
\begin{pmatrix}
b_{1,1} & b_{1,2} & b_{1,3} \\
b_{2,1} & b_{2,2} & b_{2,3} \\
b_{3,1} & b_{3,2} & b_{3,3}
\end{pmatrix}.
$$

After elimination by the first row (without division), we have the matrix

$$
\begin{pmatrix}
b_{1,1} & b_{1,2} & b_{1,3} \\
0 & b_{2,2}b_{1,1} - b_{2,1}b_{1,2} & b_{2,3}b_{1,1} - b_{2,1}b_{1,3} \\
0 & b_{3,2}b_{1,1} - b_{3,1}b_{1,2} & b_{3,3}b_{1,1} - b_{3,1}b_{1,3}
\end{pmatrix}.
$$

A second elimination gives us the matrix

$$
\begin{pmatrix}
b_{1,1} & b_{1,2} & b_{1,3} \\
0 & b_{2,2}b_{1,1} - b_{2,1}b_{1,2} & b_{2,3}b_{1,1} - b_{2,1}b_{1,3} \\
0 & 0 & \begin{aligned} b_{1,1}(b_{3,3}b_{2,2}b_{1,1} - b_{3,3}b_{2,1}b_{1,2} - b_{3,2}b_{2,3}b_{1,1} \\ + b_{3,2}b_{2,1}b_{1,3} + b_{3,1}b_{2,3}b_{1,2} - b_{3,1}b_{2,2}b_{1,3}) \end{aligned}
\end{pmatrix},
$$

and it is obvious that $b_{1,1}$ divides all the elements of the third row.

The general identity is

$$
a_{i,j}^{(k)} = \frac{1}{\left(a_{l,l}^{(l-1)}\right)^{k-l}}
\begin{vmatrix}
a_{l+1,l+1}^{(l)} & \cdots & a_{l+1,k}^{(l)} & a_{l+1,j}^{(l)} \\
\cdots & \cdots & \cdots & \cdots \\
a_{k,l+1}^{(l)} & \cdots & a_{k,k}^{(l)} & a_{k,j}^{(l)} \\
a_{i,l+1}^{(l)} & \cdots & a_{i,k}^{(l)} & a_{i,j}^{(l)}
\end{vmatrix},
$$

of which the identity we have already quoted is the special case $l = k - 1$.

2.8.3 Sparse matrices

When we are dealing with sparse matrices, Computer Algebra comes close to the methods of numerical calculation. We often use a representation where each row of the matrix is represented by a list of the non-zero elements of the row, each one stored with an indication of its column. We can also use a method of storing by columns, and there are several more complicated methods. In this case, addition is fairly simple, but multiplication is more difficult, for we want to traverse the matrix on the left row-by-row, but the matrix on the right column-by-column.

For the determinant, the general idea is Cramer's expansion, but there are several tricks for choosing the best direction for the expansion, and for using results already calculated, instead of calculating them again. Smit [1981] shows some of these techniques.

The inverse of a sparse matrix is usually dense, and therefore should not be calculated. Thus, there are three general ways of solving the system $Ax = b$ of linear equations. In the first place, there is a formula (due apparently to Laplace), which expresses the elements x_i of the solution in terms of the b_j and of the $A_{i,j}$, which are the minors of A, that is the determinants of the matrix obtained by striking out the i-th column and the j-th row of A. In fact

$$
x_i = \sum A_{i,j} b_j.
$$

These minors can be calculated by Cramer's rule, and the chances of re-using intermediate results of a calculation in subsequent calculations are good. Smit [1981] describes several tricks which improve this algorithm. But one defect which must be noted is that memory has to be used to store the reusable results.

A second way is to use Gaussian elimination (or one of the variants described for the case of a dense matrix). We can choose the rows (or columns) to be eliminated according to the number of non-zero elements they contain (and, among those which have the same number, one can try to minimise the creation of new elements, and to maximise the superimposition of non-zero elements in the addition of rows or columns). This "intelligent elimination" is fairly easy to write*, but, in general, the matrix becomes less and less sparse during these operations, and the calculating time remains at $O(n^3)$. This method was used by Coppersmith and Davenport [1985] to solve a system of 1061 equations in 739 variables, and this swell did indeed appear.

Very recently, several authors have adapted the iteration methods from numerical calculation, such as the method of Lanczos and the method of conjugate gradients, to Computer Algebra. These methods seem asymptotically more worth-while than Gaussian elimination, although, for a small problem, or even for the problem of Coppersmith and Davenport, they are less rapid. Coppersmith *et al.* [1986] discuss these methods.

2.9 REPRESENTATIONS OF SERIES

Computer Algebra is not limited to finite objects, such as polynomials. It is possible to deal with several types of infinite series. Obviously, the computer can deal with only a finite number of objects, that is, the first terms of the series.

2.9.1 Taylor series: simple method

These series are very useful for several applications, especially when it is a question of a non-linear problem, which becomes linear when we suppress some small quantities. Here one can hope that the solution can be represented by a Taylor series, and that a small number of terms suffices for the applications (e.g. numerical evaluation).

Very often these series can be calculated by the method called *successive approximation*. Let us take for example the equation $y^2 = 1 + e$, where y is an unknown and e is small, and let us look for an expression of y as a Taylor series with respect to e. Let us write y_n for the series up to the term $c_n e^n$, with $y_0 = 1$ (or -1, but we shall expand the first solution). We can calculate y_{n+1}, starting from y_n, by the following method:

$$
\begin{aligned}
1 + e &= y_{n+1}^2 + O(e^{n+2}) \\
&= \left(y_n + c_{n+1} e^{n+1}\right)^2 + O(e^{n+2}) \\
&= y_n^2 + 2y_n c_{n+1} e^{n+1} + O(e^{n+2}) \\
&= y_n^2 + 2y_0 c_{n+1} e^{n+1} + O(e^{n+2}).
\end{aligned}
$$

---

* The author has done it in less than a hundred lines in the language SCRATCHPAD-II [Jenks, 1984].

If d_{n+1} is the coefficient of e^{n+1} in $1 + e - y_n^2$, this formula implies that $c_{n+1} = d_{n+1}/2y_0$. This gives a fairly simple REDUCE program for the evaluation of y, as far as e^{10} for example:

```
ARRAY TEMP(20);
Y:=1;
FOR N:=1:10 DO <<
   COEFF(1+E-Y**2,E,TEMP);
   Y:=Y+TEMP(N)*(E**N)/2 >>;
```

This program has the disadvantage that it calculates explicitly all the terms of y^2, even those which contribute nothing to the result. REDUCE has a mechanism for avoiding these calculations: the use of WEIGHT and WTLEVEL. The details are given in the annex on REDUCE, but here we shall give the algorithm rewritten to use this mechanism.

```
WEIGHT E=1;
Y:=1;
FOR N:=1:10 DO <<
   WTLEVEL N;
   Y:=Y+(1+E-Y**2)/2 >>;
```

In this case, there are more direct methods, such as the binomial formula, or direct programming which only calculates the term in e^n of y^2, but they are rather specialised. This method of successive approximation can be applied to many other problems — Fitch [1985] gives several examples.

Series can be manipulated in the same way as the polynomials — in fact most Computer Algebra systems do not make any distinction between these two. In general, the precision (that is the highest power of e, in the case we have just dealt with) of a result is the minimum of the precisions of the data. For example, in

$$\sum_{i=0}^{n} a_i e^i + \sum_{j=0}^{m} b_j e^j = \sum_{i=0}^{\min(m,n)} (a_i + b_i)e^i,$$

the terms of the result with $i > \min(m,n)$ cannot be determined purely as a function of the data — we have to know more terms a_i or b_j than those given. In particular, if all the initial data have the same precision, the result has also the same precision — we do not get the accumulation of errors that we find in numerical calculation.

Nevertheless there is a possible loss of precision in some cases. For example, if we divide one series by another which does not begin with a term with exponent zero, such as the following series:

$$\sum_{i=1}^{n} a_i e^i \bigg/ \sum_{j=1}^{m} b_j e^j = \sum_{i=0}^{\min(m-1,n)} c_i e^i,$$

we find that there is less precision. (We note that $c_0 = a_1/b_1$ — it is more complicated to calculate the other c_i, but the method of successive approximation is often used in this calculation.)

This can also happen in the case of calculating a square root:

$$\sqrt{\sum_{i=0}^{n} a_i e^i} = \sum_{j=0}^{n} b_j e^j$$

if $a_0 \neq 0$ (in this case, $b_0 = \sqrt{a_0}$, and the other b_i can be determined, starting out from this value, by the method of successive approximation). On the other hand, if $a_0 = 0$, the series has a very different form. If $a_1 \neq 0$, the series cannot be written as a series in e, but it needs terms in $\sqrt{e}$; it is then a Puiseux series. If $a_0 = a_1 = 0$, but $a_2 \neq 0$, then the series is still a Taylor series:

$$\sqrt{\sum_{i=2}^{n} a_i e^i} = \sum_{j=1}^{n-1} b_j e^j.$$

Here there is indeed a loss of precision, for the coefficient b_n is not determined by the given coefficients a_i, but, it requires a knowledge of a_{n+1} for its determination.

We must take care to avoid these losses in precision, for simply using polynomial manipulation does not alert us to them. But the situation is much less complicated than in numerical calculation. There is no gradual loss of precision, such as that generated by numerical rounding. The circumstances which give rise to these losses are well defined, and one can program so that they are reported. If necessary, one can work to a higher precision and check that the results are the same, but this is a solution of last resort.

2.9.2 Taylor series: Norman's method

There are applications in which the losses of precision described in the last paragraph occur very frequently. Moreover, the simple method requires all the terms of the intermediate results to be calculated before the first term of the final result appears. Norman [1975] therefore suggested the following method: instead of calculating all the terms $c_0, \ldots, c_n$ of a series before using them, we state the general rule which expresses c_i in terms of these data, and leave to the system the task of calculating each c_i *at the moment when it is needed*.

For normal operations, these general rules are not very complicated, as the following table shows (an upper case letter indicates a series, and the corresponding lower case letter indicates the coefficients of the series).

$$C = A + B \quad c_i = a_i + b_i$$

$$C = A - B \quad c_i = a_i - b_i$$

$$C = A \times B \quad c_i = \sum_{j=0}^{i} a_j b_{i-j}$$

$$C = A/B \quad c_i = \frac{a_i - \sum_{j=0}^{i-1} c_j b_{i-j}}{b_0} \tag{D}$$

(The last equation only holds in the case $b_0 \neq 0$ — the general equation is a bit more complicated.) Norman has also shown that any function defined by a linear differential equation gives rise to a similar equation for the coefficients of the Taylor series.

These rules have been implemented by Norman [1975] in SCRATCH-PAD-I [Griesmer *et al.*, 1975], a system which automatically expands the values given by these rules. It is enough to ask for the value of c_5, for example, for all the necessary calculations to be done. Moreover, the system stores the values already calculated, instead of calculating them again. This last point is very important for the efficiency of this method. Let us take, for example, the case of a division (equation (D) above). The calculation of each c_i requires i additions (or subtractions), i multiplications and one division, and this gives us $(n + 1)^2$ operations for calculating $c_0, \ldots, c_n$.

But, if we suppose that the c_i are not stored, the cost is very different. The calculation of c_0 requires a division. The calculation of c_1 requires a multiplication, a subtraction and a division, but involves also the calculation of c_0, which means the cost of one addition (or subtraction), one multiplication and two divisions, which we call $[A = 1, M = 1, D = 2]$. The calculation of c_2 needs two additions/subtractions, two multiplications and one division, plus the calculation of c_0 and c_1, which costs $[A = 3, M = 3, D = 4]$, a cost which is already higher than the storage method, for c_0 has been calculated twice. The calculation of c_3 requires $[A = 3, M = 3, D = 1]$ plus the calculation of c_0, c_1, c_2, which costs $[A = 7, M = 7, D = 8]$. Similarly, the calculating cost for c_4 is $[A = 15, M = 15, D = 16]$, and the general formula for c_n is $[A = M = 2^n - 1, D = 2^n]$. The situation would have been much worse, if the data a_i and b_i had required similar calculations before being used.

But such a system of recursive calculation can be implemented in other Computer Algebra systems: Davenport [1981] has constructed a sub-system in REDUCE which expands Puiseux series (that is series with fractional exponents) of algebraic functions. The internal representation of a series is a list of the coefficients which have already been calculated, and to this is added the general rule for calculating other coefficients. For example, if **T1** and **T2** are two of these representations, the representation corresponding to **T1** ∗ **T2**, with two coefficients c_0 and c_1 already calculated, would be*:

$$(((0 \ . \ \ c_0) \ (1 \ . \ \ c_1)) \ \text{TIMES T1 T2})$$

and the command to calculate c_2 will change this structure into

$$(((0 \ . \ \ c_0) \ (1 \ . \ \ c_1) \ (2 \ . \ \ c_2)) \ \text{TIMES T1 T2}) \ ,$$

with perhaps some expansion of **T1** and **T2** if other terms of these series have been called for.

This idea of working with a infinite object of which only a finite part is expanded is quite close to the idea of *lazy evaluation* which is used in computer science. It is very easy (perhaps one hundred lines [Ehrhard, 1986]) to implement Taylor series in this way in lazy evaluation languages, such as "Lazy ML" [Mauny, 1985].

2.9.3 Other series

There are several kinds of series besides Taylor series (and its variants such as the Laurent or Puiseux series). A family of series which is very useful in several areas is that of Fourier series, that is

$$f = a_0 + \sum_{i=1}^{n} a_i \cos it + b_i \sin it.$$

The simple function $\sin t$ (or $\cos t$) represents the solution of $y'' + y = 0$, and several perturbations of this equation can be represented by Fourier series, which are often calculated by the method of successive approximation. Fitch [1985] gives some examples.

For Computer Algebra systems, based on polynomial calculation, it is more difficult to treat this series than it is to treat Taylor series. The reason is that the product of two base terms is no longer a base term : $e^i \times e^j = e^{i+j}$, but $\cos it \times \cos jt \neq \cos(i + j)t$. Of course, it is possible to re-express this product in terms of base functions, by the rewrite rules given in the section "Representation of transcendental functions", but (for the reasons given in that section) this may become quite costly. If we have to treat large series of this kind, it is more efficient to use a representation of these series, in which the multiplication can be done directly, for example a vector of the

---

* For technical reasons, Davenport uses special symbols, such as **TAY-LORTIMES** instead of **TIMES**.

coefficients of the series. There are systems in which a representation of this kind is used automatically for Fourier series — CAMAL [Fitch, 1974] is one example of this.

The reader may notice that the relation between truncation and multiplication of these series is not as clear as it was for Taylor series. If we use the notation $\lfloor \ldots \rfloor_n$ to mean that an expression has been truncated after the term of index n (for example, e^n or $\cos nt$), we see that

$$\left\lfloor \left(\sum_{i=0}^{\infty} a_i e^i \right) \left(\sum_{i=0}^{\infty} b_i e^i \right) \right\rfloor_n = \left\lfloor \left\lfloor \sum_{i=0}^{\infty} a_i e^i \right\rfloor_n \left\lfloor \sum_{i=0}^{\infty} b_i e^i \right\rfloor_n \right\rfloor_n, \qquad (T)$$

but that

$$\left\lfloor \left(\sum_{i=0}^{\infty} a_i \cos it \right) \left(\sum_{i=0}^{\infty} b_i \cos it \right) \right\rfloor_n \neq \left\lfloor \left\lfloor \sum_{i=0}^{\infty} a_i \cos it \right\rfloor_n \left\lfloor \sum_{i=0}^{\infty} b_i \cos it \right\rfloor_n \right\rfloor_n.$$
$$(F)$$

For this reason, the coefficients of our Fourier series must tend to zero in a controlled fashion, for example $a_i = O(e^i)$ where e is a quantity which is considered small. In this case, the equation (F) becomes a true equality.

The reader can easily find other series, such as Poisson series, which can be treated in the same way. The section on solutions in the form of series of linear differential equations illustrates some of these questions at greater length.

3. Polynomial simplification

3.1 SIMPLIFICATION OF POLYNOMIAL EQUATIONS

Very often, when studying some situation, we find that it is defined by a system of polynomial equations. For example, a position of the mechanical structure made up of two segments of lengths 1 and 2 respectively, and joined at one point is defined by nine variables (three points each with three co-ordinates) subject to the relations:

$$(x_1 - x_2)^2 + (y_1 - y_2)^2 + (z_1 - z_2)^2 = 1,$$
$$(x_1 - x_3)^2 + (y_1 - y_3)^2 + (z_1 - z_3)^2 = 4; \tag{1}$$

or perhaps to the relations:

$$(x_1 - x_2)^2 + (y_1 - y_2)^2 + (z_1 - z_2)^2 = 1,$$
$$(2x_1 - x_2 - x_3)(x_3 - x_2) + (2y_1 - y_2 - y_3)(y_3 - y_2) + \tag{1'}$$
$$+ (2z_1 - z_2 - z_3)(z_3 - z_2) = -3$$

(where we have replaced the second equation by the difference between the two equations); or likewise to the relations:

$$(2x_1 - x_2 - x_3)(x_3 - x_2) + (2y_1 - y_2 - y_3)(y_3 - y_2) +$$
$$+ (2z_1 - z_2 - z_3)(z_3 - z_2) = -3, \tag{1''}$$
$$(x_1 - x_3)^2 + (y_1 - y_3)^2 + (z_1 - z_3)^2 = 4;$$

or many other variants. The fundamental question of this section can be expressed thus: which finite list of relations must we use?

Every calculation about the movement of this object must take these relations into account, or, more formally, take place in the *ideal* generated by these generators.

Definition. *The ideal* generated by a family of generators consists of the set of linear combinations of these generators, with polynomial coefficients.*

Definition. *Two polynomials f and g are equivalent with respect to an ideal if their difference belongs to the ideal.*

If we regard the generators of the ideal as polynomials with value zero, this definition states that the polynomials do not differ.

Therefore, the questions for this section are:

(1) How do we define an ideal?

(2) How do we decide whether two polynomials are equivalent with respect to a given ideal?

3.1.1 Reductions of polynomials

Obviously there are several systems of generators for one ideal. We can always add any linear combination of the generators, or suppress one of them if it is a linear combination of the others. Is there a system of generators which is *simple*? Naturally this question requires us to be precise about the idea of "simple". That depends on the order over the monomials of our polynomials (see the section "Polynomials in several variables").

Let us consider polynomials in the variables $x_1, x_2, \ldots, x_n$, the coefficients of which belong to a field k (that is that we can add, multiply, divide them, etc.). Let $<$ be an order over the monomials which satisfies the following two conditions:

(a) If $a < b$, then for every monomial c, we have $ac < bc$.

(b) For all monomials a and b with $b \neq 1$, we have $a < ab$.

The three orders *lexicographic, total degree, then lexicographic* and *total degree, then inverse lexicographic* satisfy these criteria. Let us suppose that every (non-zero) polynomial is written in decreasing order (according to $<$) of its monomials: $\sum_{i=1}^{n} a_i X_i$ with $a_i \neq 0$ and $X_i > X_{i+1}$ for every i. We call X_1 the *principal monomial* and $a_1 X_1$ the *principal term* of the polynomial.

Let G be a finite set of polynomials, and $>$ a fixed order, satisfying the above conditions.

Definition. *A polynomial f is reduced with respect to G if no principal monomial of an element of G divides the principal monomial of f.*

In other words, no combination $f - h g_i$ of f and an element of G can have a principal monomial less (for the order $<$) than the principal monomial of

---

* This definition is the general definition in abstract algebra, specialised to the case of the polynomials.

f. If f is not reduced, we can subtract from it a multiple of an element of G in order to eliminate its principal monomial (and to get a new principal monomial less than the principal monomial of f); this new polynomial is equivalent to f with respect to the ideal generated by G. This process is called a *reduction* of f with respect to G. We must note that a polynomial can have several reductions with respect to G (one for each element of G whose principal monomial divides the principal monomial of f). For example, let $G = \{g_1 = x - 1, g_2 = y - 2\}$ and $f = xy$. Then there are two possible reductions of f: by g_1, which gives $f - yg_1 = -y$, and by g_2, which gives $f - xg_2 = -2x$. A polynomial f cannot have an infinite chain of reductions: we have to terminate with a reduced polynomial.

The definition of *reduced* involves the principal monomial of f, and implies that there is no linear combination $f - hg_i$ which has a principal monomial less than that of f. It is possible that there are other monomials of f which can be eliminated to make the linear combination "smaller". For example, let us suppose that the variables are x and y, subjected to the lexicographic order $y < x$, and that $G = \{y - 1\}$. The polynomial $x + y^2 + y$ is reduced, for its principal monomial is x. Nevertheless, we can eliminate the monomials y^2 and y with respect to G. This fact leads to the following definition, which is stronger than that of "reduced".

Definition. *A polynomial f is completely reduced with respect to G if no monomial of f is divisible by the principal monomial of an element of G.*

3.1.2 Standard (Gröbner) bases

Definition. *A system of generators (or basis) G of an ideal I is called a standard basis or Gröbner basis (with respect to the order $<$) if every reduction of an f of I to a reduced polynomial (with respect to G) always gives zero.*

An equivalent definition is that every f has only one reduced form with respect to G. Another, more effective, definition will be given in the next section. In the language of computer science we say that reduction with respect to G has the *Church-Rosser property*. In general, the standard bases of an ideal give us much more information about the ideal than any other bases.

For example, let us consider the ideal generated by the three polynomials

$$g_1 = x^3yz - xz^2,$$
$$g_2 = xy^2z - xyz,$$
$$g_3 = x^2y^2 - z.$$

It is obvious that $x = y = z = 0$ makes all these polynomials vanish: it is not obvious that there are other solutions. The standard basis of this ideal (with respect to the lexicographic order $x > y > z$) is formed from g_2 and g_3 and from the following three polynomials:

$$g_4 = x^2yz - z^2,$$
$$g_5 = yz^2 - z^2,$$
$$g_6 = x^2z^2 - z^3.$$

(A standard basis may very well contain more polynomials than the initial basis — we shall return to this remark at the end of the next section.) It is now obvious that there are two possibilities: either $z = 0$, or $z \neq 0$. If $z = 0$, these polynomials reduce to x^2y^2, therefore $x = 0$ or $y = 0$. If $z \neq 0$, then these polynomials reduce to $y = 1$ and $x^2 = z$. Therefore the set of common zeros of G consists of two straight lines ($x = z = 0$ and $y = z = 0$) and a parabola ($y = 1$, $x^2 = z$).

We now state some theorems on standard bases. We shall not prove them: the reader who is interested in the proofs should consult the papers by Buchberger [1970, 1976a, b, 1979, 1981, 1985].

Theorem 1. *Every ideal has a standard basis with respect to any order (which satisfies the conditions (a) and (b) above).*

Theorem 2. *Two ideals are equal if and only if they have the same reduced standard bases (with respect to the same order).*

By *reduced basis*, we mean a basis, each polynomial of which is completely reduced with respect to all the others. This restriction to reduced bases is necessary to eliminate trivialities such as $\{x - 1, (x - 1)^2\}$, which is a different basis from $\{x - 1\}$, but only because it is not reduced. In fact this theorem gives a canonical representation for the ideals (as soon as we have fixed an order).

Theorem 3. *A system of polynomial equations is inconsistent (it can never be satisfied, even if we add algebraic extensions — for example over the complex numbers* **C***) if and only if the corresponding standard basis (with respect to any order satisfying the conditions (a) and (b) above) contains a constant.*

3.1.3 Solution of a system of polynomials

Theorem 4. *A system of polynomial equations has a finite number* of solutions over* **C** *if and only if each variable appears alone (such as z^n) in one of the principal terms of the corresponding standard basis with respect to the lexicographic order.*

In this case, we can determine all the solutions by the following method. Let us suppose that the variables are $x_1, x_2, \ldots, x_n$ with $x_1 > x_2 > \ldots x_n$. The variable x_n appears alone in the principal term of a generator of the standard basis. But all the other terms of this generator are less (in the sense of $<$) than this term and therefore can contain only x_n, for we are using the lexicographic order. Thus we have a polynomial in x_n (and only one, because with two polynomials, we can always reduce one with respect to the other), which has only a finite number of roots. x_{n-1} appears by itself (to the power k, for example) in the principal term of a generator of the standard basis. But the other terms of this generator are less (in the sense of $<$) than this term, and can therefore contain only x_{n-1} (to a power less than k) and, possibly, x_n , for we are using the lexicographic order. For every possible value of x_n, we have** k possible values of x_{n-1}: the roots of this polynomial in x_{n-1}. It is possible that there are other polynomials in x_{n-1} and x_n, and that certain combinations of values of x_{n-1} and x_n do not satisfy these polynomials, and must therefore be deleted, but we are certain of having only a finite number of possibilities for x_{n-1} and x_n. We determine $x_{n-2}, \ldots, x_1$ in the same way.

It should be noted that the hypothesis is that each variable appears by itself. It cannot be replaced by the weaker hypothesis that each variable appears as principal variable, as we see from the following example. Let us consider two variables x and y, with $x > y$. Our ideal is generated by $(y - 1)x + (y - 1)$ and $y^2 - 1$, which (for reasons we shall explain later) form a standard basis, but a basis which does not satisfy the hypothesis of theorem 4, for x does not appear by itself. There are two possible solutions for y, that is 1 and -1. If $y = -1$, the other generator becomes $-2x - 2$, which vanishes when $x = -1$. But, if $y = 1$, the other generator vanishes completely, and x is not determined.

Theorem 4 has an obvious converse: an ideal is not of dimension zero if and only if there is a variable which does not appear by itself in one of the principal terms of the standard basis (with respect to the lexicographic

---

* In geometry, such an ideal is called *zero-dimensional*.
** In general. It is possible that there are multiple roots, and therefore the number of separate roots would be less than k.

order). In this case, the theory is substantially more complicated (Giusti [1984] gives some indications, but it is a subject for which algorithms are still being developed). It is easy to conjecture that the number of such variables determines the dimension of the ideal (one, if there is a curve on which all the polynomials vanish; two, if it is a question of a surface etc.), but this conjecture is false. Let us take as an example the following two ideals:

$$I_1 = \{x^2 - 1, (x - 1)y, (x + 1)z\}$$
$$I_2 = \{x^2 - 1, (x - 1)y, (x - 1)z\}.$$

Their standard bases with respect to the lexicographic order $x > y > z$ are

$$I_1 = \{x^2 - 1, (x - 1)y, (x + 1)z, yz\}$$
$$I_2 = \{x^2 - 1, (x - 1)y, (x - 1)z\}.$$

In both, x appears by itself, but neither y nor z does. Now I_1 has dimension one, for its solutions are the two straight lines $x = 1$, $z = 0$ and $x = -1$, $y = 0$, whereas I_2 has dimension 2, for its solutions are the plane $x = 1$ and the isolated point $x = -1$, $y = z = 0$.

3.1.4 Buchberger's algorithm

Theorem 1 of the last section but one tells us that every ideal has a standard basis — but how do we calculate it? Similarly, how can we decide whether a basis is standard or not? In this section we shall explain Buchberger's algorithm [1970], with which we can solve these problems. We suppose that we have chosen once and for all an order on the monomials, which satisfies conditions (a) and (b) of the section "Reductions of polynomials".

Definition. *Let f and g be two non-zero polynomials, with principal terms f_p and g_p. Let h be the l.c.m. of f_p and g_p. The S-polynomial of f and g, $S(f, g)$, is defined by*

$$S(f, g) = \frac{h}{f_p} f - \frac{h}{g_p} g.$$

The l.c.m. of two terms or monomials is simply the product of all the variables, each to a power which is the maximum of its powers in the two monomials. h/g_p and h/f_p are monomials, therefore $S(f, g)$ is a linear combination with polynomial coefficients of f and g, and belongs to the ideal generated by f and g. Moreover, the principal terms of the two components of $S(f, g)$ are equal (to h), and therefore cancel each other. We note too that $S(f, f) = 0$ and that $S(g, f) = -S(f, g)$.

Theorem 5. *A basis G is a standard basis if, and only if, for every pair of polynomials f and g of G, $S(f, g)$ reduces to zero with respect to G.*

This theorem gives us a criterion for deciding whether a basis is standard or not — it is enough to calculate all the S-polynomials and to check that they do reduce to zero. But if we do not have a standard basis, it is precisely because one of these S-polynomials (say $S(f, g)$) does not reduce to zero. Then, as its reduction is a linear combination of the elements of G, we can add it to G without changing the ideal generated. After this addition, $S(f, g)$ reduces to zero, but there are new S-polynomials to be considered. It is a remarkable fact, which we again owe to Buchberger [1970], that this process always comes to an end (and therefore gives a standard basis of the ideal).

Let us apply this algorithm to the example $\{g_1, g_2, g_3\}$ of the previous section. The S-polynomials to be considered are $S(g_1, g_2)$, $S(g_1, g_3)$ and $S(g_2, g_3)$. The principal terms of $g_2 = xy^2z - xyz$ and $g_3 = x^2y^2 - z$ are xy^2z and x^2y^2, whose l.c.m. is x^2y^2z. Therefore

$$S(g_2, g_3) = xg_2 - zg_3 = (x^2y^2z - x^2yz) - (x^2y^2z - z^2) = -x^2yz + z^2.$$

This polynomial is non-zero and reduced with respect to G, and therefore G is not a standard basis. Therefore we can add this polynomial (or, to make the calculations more readable, its negative) to G — call it g_4. Now G consists of

$$g_1 = x^3yz - xz^2,$$
$$g_2 = xy^2z - xyz,$$
$$g_3 = x^2y^2 - z,$$
$$g_4 = x^2yz - z^2,$$

and the S-polynomials to be considered are $S(g_1, g_2)$, $S(g_1, g_3)$, $S(g_1, g_4)$, $S(g_2, g_4)$ and $S(g_3, g_4)$. Fortunately, we can make a simplification, by observing that $g_1 = xg_4$, and therefore the ideal generated by G does not change if we suppress g_1. This simplification leaves us with two S-polynomials to consider: $S(g_2, g_4)$ and $S(g_3, g_4)$.

$$S(g_2, g_4) = xg_2 - yg_4 = -x^2yz + yz^2,$$

and this last polynomial can be reduced (by adding g_4), which gives us $yz^2 - z^2$. As it is not zero, the basis is not standard, and we must enlarge G by adding this new generator, which we call g_5. Now G consists of

$$g_2 = xy^2z - xyz,$$
$$g_3 = x^2y^2 - z,$$
$$g_4 = x^2yz - z^2,$$
$$g_5 = yz^2 - z^2,$$

and the S-polynomials to be considered are $S(g_3, g_4)$, $S(g_2, g_5)$, $S(g_3, g_5)$ and $S(g_4, g_5)$.

$$S(g_3, g_4) = zg_3 - yg_4 = -z^2 + yz^2,$$

and this last one can be reduced to zero (by adding g_5) (in fact, this reduction follows from Buchberger's third criterion, which we shall describe later).

$$S(g_2, g_5) = zg_2 - xyg_5 = -xyz^2 + xyz^2 = 0.$$

$$S(g_4, g_5) = zg_4 - x^2g_5 = -z^3 + x^2z^2 = x^2z^2 - z^3,$$

where the last rewriting arranges the monomials in decreasing order (with respect to $<$). This polynomial is already reduced with respect to G, G is therefore not a standard basis, and we must add this new polynomial to G — let us call it g_6. Now G consists of

$$g_2 = xy^2z - xyz,$$
$$g_3 = x^2y^2 - z,$$
$$g_4 = x^2yz - z^2,$$
$$g_5 = yz^2 - z^2,$$
$$g_6 = x^2z^2 - z^3,$$

and the S-polynomials to be considered are $S(g_3, g_5)$, $S(g_2, g_6)$, $S(g_3, g_6)$, $S(g_4, g_6)$ and $S(g_5, g_6)$. The reader can check that G reduces all these S-polynomials to zero, and that G is therefore a standard basis of the ideal.

We must note that this algorithm can benefit from several optimisations. Buchberger [1979] has given a criterion (*Buchberger's third criterion*) which eliminates several of the S-polynomials studied. He proves that, if there is an element h of this basis, such that the principal monomial of h divides the l.c.m. of the principal monomials of f and g, and if we have already considered $S(f, h)$ and $S(h, g)$, then it is not necessary to consider $S(f, g)$, for it should reduce to zero.

We have not mentioned the choice of the S-polynomial to be studied. In general, there are several S-polynomials to be considered, even after applying Buchberger's criterion. The algorithm gives the same result for every choice, but the calculating time may be very different. Buchberger [1979] claims that the choice of an S-polynomial such that the l.c.m. of the principal monomials is of minimal total degree (amongst all the S-polynomials to be considered) is good, and works well with his criterion. But this may still leave several possibilities open. There are many heuristics, of which it is claimed that they choose a "good" S-polynomial from the list of S-polynomials, but this is an active research topic.

Nevertheless, the fact that we have an algorithm does not mean that every problem can be solved easily. Although Buchberger has proved that his algorithm terminates, he has not given any limit for the calculating time or for the number of polynomials in the standard basis. In fact, these problems are very difficult, and are being actively studied. Mayr and Mayer [1982] have shown that calculating a standard basis requires, in general, a memory space exponential in n, the number of variables. In practice, there are problems which have been solved very easily by this algorithm [Gebauer and Kredel, 1984], but also problems which use several megabytes of memory without being solved*.

The case of one variable is trivial (it is discussed in the next section). For the case of two variables, Buchberger [1983] and Lazard [1983] were able to prove that, if all the data are of total degree bounded by d, the total degrees of a standard basis are bounded by $2d - 1$ if the order is a "total degree" order, and always bounded by d^2. Moreover, the number of polynomials in the standard basis is bounded by $k + 1$, where k is the minimum of the total degrees of the principal monomials of all the data. All these limits are best possible.

The case of more than two variables is more complicated. Mora has given [Lazard, 1983] an example of a seemingly small problem,

$$\{x^{k+1} - y^{k-1}zt, xz^{k-1} - y^k, x^ky - z^kt\}$$

with the order total degree, then inverse lexicographic $x > z > y > t$, but such that the standard basis contains the polynomial $y^{k^2+1} - z^{k^2}t$, of degree almost the square of the degree of the data.

3.1.5 Relationship to other methods

Buchberger's algorithm is, in fact, a combination of several well-known ideas. For the case of one variable and two polynomials, it is equivalent to Euclid's algorithm for the g.c.d. of these polynomials. Every S-polynomial is the polynomial of highest degree less (a multiple of) the other, and after increasing the basis by this S-polynomial, the polynomial of higher degree can be dropped, for it reduces to zero with respect to the others. The extension to several polynomials is closely linked to the corresponding generalisation of Euclid's algorithm.

For the case of several variables and linear polynomials, Buchberger's algorithm corresponds to Gaussian elimination, for $S(f_1, f_2)$ (supposing that f_1 and f_2 contain the same principal variable —if not $S(f_1, f_2) = 0$) is a

---

* Note added in proof: M. Giusti tells us that he and Mora have just found a general limit.

linear combination (with constant coefficients) of f_1 and f_2 which eliminates the principal variable. f_2 reduces to zero with respect to f_1 and $S(f_1, f_2)$, and can therefore be cancelled. Thus we eliminate the principal variable of all the polynomials except f_1, and then we continue with the next variable. We end up (if the system is not singular) with a triangular system, that is one which satisfies the hypothesis of theorem 4, and the algorithm indicated in the proof of it is equivalent to the use of back-substitution to determine the solution of a triangular linear system.

For general systems, there is another method for finding the solutions, that of *repeated elimination* [Moses, 1966]. For example, for the polynomials g_1, g_2 and g_3 which we have used as an example, we can eliminate x from these polynomials (using the resultant with respect to x). This gives us polynomials of the form

$$(y - 1)^a y^b z^c$$

(a, b and c being integers). These polynomials seem to suggest that there are three families of solutions: $y = 1$ with z and x related (in fact by $x^2 = z$, as we already know); $y = 0$ with z and x related (in fact the relationship is a little special — $z = 0$) and $z = 0$ with y and x related (here the relationship is $x = 0$).

In this case, elimination has given the same results as Buchberger's algorithm, but, in general, elimination may give, not only the true solutions, but also some *parasitic solutions*, that is solutions of the reduced system which are not solutions of the given system. Let us take a fairly simple example: that of the equations

$$(y - 1)x + (y - 1) \quad \text{and} \quad (y + 1)x.$$

If we eliminate x from the second equation, we find the system

$$(y - 1)x + (y - 1) \quad \text{and} \quad y^2 - 1,$$

of which we already know that the solutions are the point $y = x = -1$ and the line $y = 1$ with x undetermined. On the contrary, if we eliminate x from the first equation, we find the system

$$(y + 1)x \quad \text{and} \quad y^2 - 1,$$

the solutions of which are the line $y = -1$ with x undetermined, and the point $y = 1$ with $x = 0$. Thus, we see that elimination may even give inconsistent results, depending on the order of elimination. To make the elimination correct, we must check that all the solutions satisfy all the given equations. If the equations are not satisfied, it is still possible that a subset of the solution holds, such as the subset $y = x = -1$ of the "solution" $y = -1$ and x undetermined, which we found before.

The corresponding standard basis is

$$2x - y + 1 \quad \text{and} \quad y^2 - 1$$

the solutions of which are finite (theorem 4), and are given explicitly by $y = \pm 1$ with the corresponding values of x: $x = 0$ (when $y = 1$) and $x = -1$ (when $y = -1$).

3.2 SIMPLIFICATION OF REAL POLYNOMIAL SYSTEMS

In this section, we shall study a problem which may seem very similar to the problem in the last section. The latter dealt with the simplification of a system of polynomial equations over an arbitrary field, and gives an algorithm for expressing the equations in canonical form, for determining whether two systems are equivalent etc. Now we shall consider the same questions, but restricting ourselves to *real* values. This is, indeed, the setting needed for most problems arising from robotics, C.A.D. (Computer Aided Design) etc. In fact, we shall see that these two situations are quite different.

Real systems differ from the other systems in two ways:

(1) one can add *inequalities*, something which has no meaning for complex numbers;

(2) the dimension of the space of solutions is not as obvious as before, given some equations (or even a single equation).

Let us enlarge upon this last point. Suppose we have two variables, x and y. If we add a non-trivial equation, we have in the case of complex numbers, a one-dimensional space of solutions. With real numbers the case is more complicated.

- The equation $x^2 + y^2 = 1$ gives a one-dimensional space, that is the circle.
- The equation $x^2 + y^2 = 0$ gives a zero-dimensional space, that is the point $x = y = 0$.
- The equation $x^2 + y^2 = -1$ gives an empty space, for this equation has no real solution.

As we said, research on the problems of the previous section is continuing, and there are a lot of important questions still unsolved. This is even truer of our present problem; so we shall only give a brief survey of the known methods and algorithms.

3.2.1 The case of $\mathbf{R}^1$

This is relatively simple. We take several polynomial equalities and inequalities in one variable, and we ask what is the number of real solutions. We shall always study the case where all the coefficients are rational numbers. It suffices, at least in principle, to study the case of a system of the form

$$p_1(x) = \cdots = p_k(x) = 0; \quad q_1(x) > 0; \ldots; \quad q_k(x) > 0,$$

since the solutions to a system which contains an inequality of the form $q_i(x) \geq 0$ are the union of those of the system with $q_i(x) = 0$ and those of the system with $q_i(x) > 0$. Moreover, it is possible to replace the p_i by their g.c.d. (written p), for a common root is necessarily a root of the g.c.d.

To solve this system, it is enough to know about all the roots of p and of the q_i, where the word "know" has a special meaning.

Definition. *A root α of a polynomial p is said to be isolated if two rational numbers a and b are given such that $a \leq \alpha \leq b$ and p has only one root in the interval $[a, b]$. This interval is called the isolating interval of α.*

A degenerate case of this definition is the case of a rational root, for which it suffices to choose $a = \alpha = b$ (but this choice is not obligatory).

Proposition. *Given an isolated root of a polynomial, we can reduce the size of the isolating interval, so that it is as small as desired.*

Proof. For simple roots, the procedure is straightforward. In the cases $a = \alpha$ (that is $p(a) = 0$) or $\alpha = b$, the interval can be reduced to a point. In the other cases, we calculate $c = \frac{1}{2}(a + b)$, and we determine the signs of $p(a)$, $p(c)$ and $p(b)$. Those of $p(a)$ and of $p(b)$ are necessarily different, and the sign of $p(c)$ is equal to one of them. We reject the interval $[a, c]$ (or $(c, b]$) on which the sign does not change, and keep the other — $[c, b]$ (or $[a, c]$), in which the root is to be found. We have thus divided the size of the interval by two, and we can go on until the interval is sufficiently small.

If we want to treat polynomials with multiple roots, we have to use the sign of the square-free part of p, that is $p/\gcd(p, p')$.

With the help of these ideas, we can construct the following algorithm for solving our problem.

[1] Isolate all the roots of p and of the q_i.

[2] Reject those roots of p which are also roots of the q_i, for they do not satisfy $q_i(\alpha) > 0$. In other words:

$$p_1 := p/\gcd(p, p');$$
$$\textbf{for all } i \textbf{ do } p_1 := p_1/\gcd(p_1, q_i);$$

and keep only those roots of p where p_1 changes sign in the isolating interval.

[3] Reduce the intervals so that each interval defining a root of p_1 does not contain a root of one of the q_i, i.e. so that the isolating intervals of p and of the q_i are disjoint.

[4] For each root α of p_1, isolated between a and b, check that all the q_i are positive. It is enough to check that $q_i(a) > 0$, for we have ensured that q_i has no root between a and α, and cannot therefore change sign.

This algorithm is purely rational and does not require any calculation with floating-point numbers. It appeals to an isolating algorithm, which we shall now describe.

3.2.1.1 *Isolation of roots*

In this section, we shall consider a polynomial $p(x) = \sum_{i=0}^{n} a_i x^i$, with integer coefficients. This latter limitation is not really a limitation, for the roots of a polynomial do not change if we clear denominators. Without loss of generality, we may suppose that p has no multiple factors (that is, that p and p' are relatively prime), for the factors do not change the roots, but only their multiplicities. This is a problem which has been studied by many famous mathematicians, for example, Descartes in the seventeenth century, Fourier, Sturm and Hermite in the nineteenth century, and by Computer Algebraists since the birth of the subject. Amongst recent papers, we mention those by Collins and Loos [1982] and by Davenport [1985b]: we give a summary of the latter (omitting the proofs).

Definition. *The Sturm sequence associated with the polynomial p is a sequence of polynomials with rational coefficients $p_0(x)$, $p_1(x)$, ..., $p_k(x)$ (where $k \leq n$ and p_k is constant), defined by the following equations:*

$$p_0(x) = p(x),$$
$$p_1(x) = p'(x), \tag{1}$$
$$p_i(x) = -\text{remainder}(p_{i-2}(x), p_{i-1}(x))$$

where "remainder" means the remainder from the division of one polynomial of $\mathbf{Q}[x]$ by another.

In fact, we only need the sign of an evaluation of the elements of a Sturm sequence; so it is appropriate to clear its denominators, and to treat the sequence as a sequence of elements of $\mathbf{Z}[x]$. This is very similar to the calculation of a g.c.d., and the elements of a Sturm sequence are (to within a sign — a subtlety which it is easy to overlook) the successive terms of the application of Euclid's algorithm to p and p'; and this is why we always end up with a constant, for p and p' were supposed relatively prime. Sturm sequences can be calculated by all the methods for calculating the g.c.d.:

- naïvely (but we strongly recommend that this is not used — see the section "The g.c.d.s" of the previous chapter for an example of the growth of the integers involved);
- by the method of sub-resultants (see the same section), but we have to make certain that the factors suppressed in this method do not change the sign of the polynomial;
- by the modular method (described in the next chapter), but here there is a serious problem with the treatment of signs and this is quite hard to solve.

Sturm sequences are interesting because of their relationship with the real roots of p, and this relationship is clarified in the following definition and theorem.

Definition. *Let y be a real number which is not a root of p. The variation at the point y of the Sturm sequence associated with p, written $V(y)$, is the number of variations of sign in the sequence $p_0(y)$, $p_1(y)$,...,$p_k(y)$.*

For example, if the values of the elements of the sequence are -1, 2, 1, 0 and -2, the variation is 2, for the sign changes between -1 and 2, and between 1 and -2 (the zeros are always ignored).

Sturm's Theorem. *If a and b are two points, which are not real roots of p, such that $a < b$, then the number of real roots of p in the interval (a,b) is $V(a) - V(b)$.*

It is possible to take the values ∞ and $-\infty$ for a and b, by taking for $V(\infty)$ the number of variations of sign between the leading coefficients of the elements of the Sturm sequence, and for $V(-\infty)$ the same number after changing the signs of the leading coefficients of the elements of odd degree. In particular, the number of real roots is a function of the signs of the leading coefficients.

For this theorem to give us an algorithm for finding the real roots of a polynomial, we have to bound the roots of this polynomial, so that we can start our search from a finite interval. There are three such bounds, which we cite here.

Proposition 1 [Cauchy, 1829]. *Let α be a root of p. Then*

$$|\alpha| \leq 1 + \max\left(\left|\frac{a_{n-1}}{a_n}\right|, \left|\frac{a_{n-2}}{a_n}\right|, \ldots, \left|\frac{a_0}{a_n}\right|\right).$$

This bound is invariant if we multiply the polynomial by a constant, but it behaves badly under the transformation $x \to x/2$, which only changes the roots by a factor 2, but may change the bound by a factor 2^n. The next two bounds do not have this defect.

Proposition 2 [Cauchy, 1829]. *Let α be a root of p. Then*

$$|\alpha| \le \max\left(\left|\frac{na_{n-1}}{a_n}\right|, \left|\frac{na_{n-2}}{a_n}\right|^{1/2}, \left|\frac{na_{n-3}}{a_n}\right|^{1/3} \cdots, \left|\frac{na_0}{a_n}\right|^{1/n}\right).$$

Proposition 3 [Knuth, 1981]. *Let α be a root of p. Then*

$$|\alpha| \le 2\max\left(\left|\frac{a_{n-1}}{a_n}\right|, \left|\frac{a_{n-2}}{a_n}\right|^{1/2}, \left|\frac{a_{n-3}}{a_n}\right|^{1/3} \cdots, \left|\frac{a_0}{a_n}\right|^{1/n}\right).$$

Each of these propositions also gives us a bound for the minimal absolute value of a root of a polynomial (supposing that the constant coefficient is not zero) — we replace x by $1/x$ in the polynomial, or, in other words, we look for the largest root of $a_0 x^n + a_1 x^{n-1} + \cdots + a_{n-1}x + a_n$, the reciprocal of which is the smallest root of $a_n x^n + a_{n-1}x^{n-1} + \cdots + a_1 x + a_0$.

These bounds give us the isolating algorithm displayed below. *Result* is a list of intervals (that is of pairs of rational numbers), each of which contains only one root. *Work* is a list of structures, containing two rational numbers defining an interval with several roots, and the values of the variation of the Sturm sequence at these two points. We suppose that *Maximal_Bound* and *Minimal_Bound* return rational numbers. In this case, all the numbers manipulated by this algorithm are rational, and the calculations can be carried out without any loss of accuracy.

We stress the reliability of this algorithm, because the roots of a polynomial, especially the real roots, are very unstable as a function of the coefficients. A very good example of this instability is given by Wilkinson [1959], who considers the polynomial

$$W(x) = (x + 1)(x + 2)\cdots(x + 20) = x^{20} + 210x^{19} + \cdots + 20!.$$

This polynomial looks completely normal, the roots are well separated from one another etc. The leading coefficients of the elements of its Sturm sequence are all positive, and we deduce that $V(\infty) = 0$, $V(-\infty) = 20$, and that the polynomial has 20 real roots. Let us consider, as Wilkinson did, a very small perturbation of this polynomial, that is

$$\widehat{W}(x) = W(x) + 2^{-23}x^{19} = x^{20} + (210 + 2^{-23})x^{19} + \cdots + 20!$$

(the number 2^{-23} is chosen to change the last bit out of 32 in the coefficient of x^{19}). It seems "obvious" that $\widehat{W}$ and W are so close that $\widehat{W}$ must also have 20 real roots, but that is completely false. In fact, $\widehat{W}$ has only ten real roots, and the imaginary parts of the other roots are fairly big — between

Isolating algorithm for real roots

$S := \{p_0, \ldots, p_k\} := Sturm(p);$
$N := Variations(S, -\infty) - Variations(S, \infty);$
if $N = 0$ **then return** $\emptyset$;
if $N = 1$ **then return** $[-\infty, \infty]$;
$M := Maximal\_Bound(p);$
$Result := \emptyset;$
$Work := \{[-M, M, Variations(S, -\infty), Variations(S, \infty)]\};$
while $Work \neq \emptyset$ **do**;
$\qquad [a, b, V_a, V_b] := element\_of(Work);$
$\qquad Work := Work \setminus \{[a, b, V_a, V_b]\};$
$\qquad c := \frac{1}{2}(a + b);$
$\qquad$ **if** $p(c) = 0$ **then** $Result := Result \cup \{[c, c]\};$
$\qquad\qquad\qquad\qquad\qquad M := Minimal\_Bound(subst(x = x - c, p)/x);$
$\qquad\qquad\qquad\qquad\qquad V_{c+} := Variation(c + M);$
$\qquad\qquad\qquad\qquad\qquad V_{c-} := V_{c+} + 1;$
$\qquad\qquad\qquad\qquad\qquad$ **if** $V_a = V_{c-} + 1$ **then**
$\qquad\qquad\qquad\qquad\qquad\qquad Result := Result \cup \{[a, c - M]\};$
$\qquad\qquad\qquad\qquad\qquad$ **if** $V_a > V_{c-} + 1$ **then**
$\qquad\qquad\qquad\qquad\qquad\qquad Work := Work \cup \{[a, c - M, V_a, V_{c-}]\};$
$\qquad\qquad\qquad\qquad\qquad$ **if** $V_b = V_{c+} - 1$ **then**
$\qquad\qquad\qquad\qquad\qquad\qquad Result := Result \cup \{[c + M, b]\};$
$\qquad\qquad\qquad\qquad\qquad$ **if** $V_b < V_{c+} - 1$ **then**
$\qquad\qquad\qquad\qquad\qquad\qquad Work := Work \cup \{[c + M, b, V_{c+}, V_b]\};$
$\qquad\qquad\qquad$ **else** $V_c := Variation(S, c);$
$\qquad\qquad\qquad\qquad\qquad$ **if** $V_a = V_c + 1$ **then**
$\qquad\qquad\qquad\qquad\qquad\qquad Result := Result \cup \{[a, c]\};$
$\qquad\qquad\qquad\qquad\qquad$ **if** $V_a > V_c + 1$ **then**
$\qquad\qquad\qquad\qquad\qquad\qquad Work := Work \cup \{[a, c, V_a, V_c]\};$
$\qquad\qquad\qquad\qquad\qquad$ **if** $V_b = V_c - 1$ **then**
$\qquad\qquad\qquad\qquad\qquad\qquad Result := Result \cup \{[c, b]\};$
$\qquad\qquad\qquad\qquad\qquad$ **if** $V_b < V_c - 1$ **then**
$\qquad\qquad\qquad\qquad\qquad\qquad Work := Work \cup \{[c, b, V_c, V_b]\};$
$\quad$ **return** $Result;$

0.8 and 3. The signs of the leading coefficients of the Sturm sequence of $\widehat{W}$ become negative from the eighth polynomial on, and $V(\infty) = 15$, whereas $V(-\infty) = 5$.

This reliability is all the more important because the interesting questions in mechanics, C.A.D. or robotics are often about the existence or not of real roots, or the value of a parameter, for which the roots of interest begin to exist. It looks as if Computer Algebra is the only tool capable of answering these questions.

However, we must be aware that these calculations can be fairly expensive. For example, we have already seen that the resultants and the coefficients which appear in the Sturm sequence may be large. The fact that the roots of a polynomial may be very close to one another implies that many bisections may be needed before separating two roots, and that the numerators and denominators which occur in the rational numbers may become quite big. All this may considerably increase the running time for the algorithm in question.

Theorem [Davenport, 1985b]. *The running time of this algorithm is bounded by* $O(n^6(\log n + \log \sum a_i^2)^3)$.

This bound is somewhat pessimistic, and it seems that the average time is more like $O(n^4)$ [Heindel, 1971]. Other methods, which may be more efficient, are described by Collins and Loos [1982].

3.2.1.2 *Real algebraic numbers*

The previous chapter dealt with the representation of algebraic numbers from a purely algebraic point of view and we did not distinguish between the different roots of an irreducible polynomial. That point of view is appropriate for many applications, such as integration. But we now have available the tools necessary for dealing with algebraic numbers in a more numerical way.

Definition. *A representation of a real algebraic number consists of: a square-free polynomial p with integral coefficients; and an isolating interval* $[a, b]$ *of one of its roots.*

We have not insisted on the polynomial being irreducible, for we have all the information needed to answer questions such as "Is this algebraic number a root of this other polynomial q?". In fact, the solution of this question is fairly simple — "yes" if and only if the g.c.d. of the two polynomials has a root in the interval $[a, b]$, and one can test this by checking whether the g.c.d. changes sign between a and b (the g.c.d. cannot have multiple roots).

As we have already said, by starting from such a representation, we can produce rational approximations (and therefore approximations in floating-point numbers) of arbitrary precision.

3.2.2 The general case — (some definitions)

Before we can deal with this case, we need several definitions which are, in some sense, generalisations of the ideas of "root" and of "interval".

Definition. *A semi-algebraic component of* $\mathbf{R}^n$ *is a set of points satisfying a family of polynomial equalities and inequalities:*

$$\{(x_1,\ldots,x_n) : p_1(x_1,\ldots,x_n) = \cdots = p_k(x_1,\ldots,x_n) = 0;$$
$$q_1(x_1,\ldots,x_n) > 0;\ldots;q_k(x_1,\ldots,x_n) > 0\}.$$

This is, obviously, the natural generalisation of the situation in $\mathbf{R}^1$, which we have just looked at. In fact, it is more natural to look, not only at the components, but also at the objects constructed by set operations starting from components.

Definition. *A semi-algebraic variety is either a semi-algebraic component, or one of the sets* $A \cup B$, $A \cap B$ *and* $A \setminus B$, *where* A *and* B *are two semi-algebraic varieties.*

This definition characterises the sets of $\mathbf{R}^n$ which can be described in terms of polynomial equalities and inequalities, for example

$$\{(x,y,z) : x^2 + y^2 + z^2 > 0 \text{ or } (x \neq 0 \text{ and } y^2 - z \leq 0)\}$$

can be written as a semi-algebraic variety in the form

$$A \cup ((B \cup C) \cap (D \cup E)),$$

where

$$A = \{(x,y,z) : x^2 + y^2 + z^2 > 0\}$$
$$B = \{(x,y,z) : x > 0\}$$

$$C = \{(x,y,z) : x < 0\}$$
$$D = \{(x,y,z) : y^2 - z < 0\}$$
$$E = \{(x,y,z) : y^2 - z = 0\}.$$

We can test whether a point (with rational or algebraic coefficients, in the sense defined in the previous section) belongs to such a variety, by checking whether the point belongs to each component, and by applying the Boolean laws corresponding to the construction of the variety. We can do this more economically: if the variety is of the form $A \cup B$, there is no need to test the components of B if the point belongs to A.

But if we start from such a description, it is very hard to understand the structure of a semi-algebraic variety, even if we can determine whether some points belong to the variety. Here are some questions which can be asked about a variety:

(1) Is it empty?
(2) What is its dimension?
(3) Is it connected?

3.2.2.1 *Decomposition of* $\mathbf{R}^n$

In this section, we shall look at a method for decomposing $\mathbf{R}^n$ which allows us to analyse the structure of a semi-algebraic variety.

Definition. *A decomposition of $\mathbf{R}^n$ is the representation of $\mathbf{R}^n$ as the union of a finite number of disjoint and connected semi-algebraic components.*

By requiring the components to be disjoint and connected, a decomposition is already rather more manageable than an arbitrary semi-algebraic variety. It is often useful to know one point in each component of a decomposition.

Definition. *A decomposition is pointed if, for every component, a point with rational or algebraic coefficients belonging to this component is associated with it.*

The points given in the components form a set of representatives of $\mathbf{R}^n$ with a point in each component. Before we can use such a set, we must know that it really is representative.

Definition. *A decomposition of $\mathbf{R}^n$ is invariant for the signs of a family of polynomials $r_i(x_1, \ldots, x_n)$ if, over each component of the decomposition, each polynomial is:*
- *always positive* **or**
- *always negative* **or**
- *always zero.*

For example, the decomposition of $\mathbf{R}^1$ as

$$(-\infty, -2) \cup \{-2\} \cup (-2, 1) \cup \{1\} \cup (1, \infty)$$

(which can be written more formally as

$$\{x : -x - 2 > 0\} \cup \{x : x + 2 = 0\} \cup \{x : x + 2 > 0 \text{ and } 1 - x > 0\}$$
$$\cup \{x : x - 1 = 0\} \cup \{x : x - 1 > 0\}$$

according to the formal definition of a decomposition) is invariant for the sign of the polynomial $(x + 2)(x - 1)$.

Proposition. *Let V be a semi-algebraic variety, and D a decomposition of $\mathbf{R}^n$ invariant for the signs of any polynomials occurring in the definition of V (we shall abbreviate this to "invariant for the signs of V"). For every component C of D, $C \cap V = \emptyset$ or $C \subset V$.*

In fact, a pointed decomposition which is invariant for the signs of V gives us a lot of information about the variety V. Starting from such a decompostion, we can test whether the variety is empty or not by testing whether each of the points marking the decomposition belongs to the variety or not. If the variety contains a point of $\mathbf{R}^n$, then the proposition implies that it contains the whole of the component it is in, and therefore also the point which marks it.

3.2.3 Cylindrical decompositions

Although the idea of decomposition is useful, it is not sufficiently constructive to enable us to compute with any decomposition. In fact, the previous section is not really part of Computer Algebra, because the ideas in it are not completely algorithmic. In this section we shall define a type of decomposition which is much more constructive, and we shall sketch an algorithm for calculating it. We suppose that we have chosen a system of co-ordinates $x_1, \ldots, x_n$ of $\mathbf{R}^n$.

Definition. *A decomposition D of $\mathbf{R}^n$, that is $\mathbf{R}^n = E_1 \cup \cdots \cup E_N$, is cylindrical if $n = 0$ (the trivial case) or if:*
(a) *$\mathbf{R}^{n-1}$ has a cylindrical decomposition D', which we can write $\mathbf{R}^{n-1} = F_1 \cup \cdots \cup F_m$ and*
(b) *For each component E_i of D, there is a component F_j of D' such that E_i is written in one of the following forms*

$$\{(x_1, \ldots, x_n) : (x_1, \ldots, x_{n-1}) \in F_j \text{ and } x_n < f_k(x_1, \ldots, x_{n-1})\} \ (1)$$

$$\{(x_1, \ldots, x_n) : (x_1, \ldots, x_{n-1}) \in F_j \text{ and } x_n = f_k(x_1, \ldots, x_{n-1})\} \ (2)$$

$$\{(x_1, \ldots, x_n) : (x_1, \ldots, x_{n-1}) \in F_j \text{ and } f_k(x_1, \ldots, x_{n-1}) < x_n$$
$$\text{and } x_n < f_{k'}(x_1, \ldots, x_{n-1})\} \ (3)$$

$$\{(x_1, \ldots, x_n) : (x_1, \ldots, x_{n-1}) \in F_j \text{ and } x_n > f_k(x_1, \ldots, x_{n-1})\} \ (4)$$

where the f_k are the solutions of polynomial equations with rational coefficients, for example "$f_k(x_1, \ldots, x_{n-1})$ is the third real root of the polynomial $p(x_1, \ldots, x_{n-1}, z)$, considered as a polynomial in z".

The idea behind this definition, introduced by Collins [1975], is quite simple — $\mathbf{R}^{n-1}$ is divided according to the decomposition D', and above each component F_j we consider the cylinder* of all the points $(x_1, \ldots, x_n)$ with $(x_1, \ldots, x_{n-1})$ belonging to F_j. We require this cylinder to be intersected by a finite number N of surfaces, which are single-valued with respect to $(x_1, \ldots, x_{n-1})$, i.e. which can be written in the form $x_n = f_k(x_1, \ldots, x_{n-1})$. Moreover, we require that these surfaces do not touch, and *a fortiori* that they do not cross. So it is possible to arrange the surfaces in increasing order of x_n for a point $(x_1, \ldots, x_{n-1})$ of a given F_j, and this order is the same for every point of F_j. Thus the cylinder is divided into

---

* More precisely, the generalised cylinder. If F_j is a disc, we have a true cylinder.

N surfaces of type (2), of the same dimension as that of F_j, $N - 1$ "slices" between two surfaces given by equations of type (3) and of dimension equal to that of F_j plus one, and two semi-infinite "slices", given by equations of type (1) and (4), likewise of dimension one plus that of F_j.

A cylindrical decomposition of $\mathbf{R}^1$, invariant for the signs of a family of polynomials, is quite easy to calculate — we have to know about (in the sense of isolating them) all the roots of the polynomials, and the decomposition is made up of these roots and of the intervals between them.

Theorem [Collins, 1975]. *There is an algorithm which calculates a cylindrical decomposition of $\mathbf{R}^n$ invariant for the signs of a family P of polynomials with integer coefficients. If the family P contains m polynomials, of degree less than or equal to d and with coefficients bounded by 2^H (length bounded by H), the time taken to finish this algorithm is bounded by*

$$(2d)^{2^{2n+8}} m^{2^{n+6}} H^3. \tag{5}$$

The principle behind this algorithm is quite simple: it is based on the following plan for $n > 1$, and we have already seen how to solve this problem in the case $n = 1$.

[1] Calculate a family Q of polynomials in the variables $x_1, \ldots, x_{n-1}$, such that a decomposition of $\mathbf{R}^{n-1}$ invariant for their signs can serve as a basis for the decomposition of $\mathbf{R}^n$ invariant for the signs of P.

[2] Calculate, by recursion, a decomposition D' of $\mathbf{R}^{n-1}$ invariant for the signs of Q.

[3] For each component F of D', calculate the decomposition of the cylinder above F induced by the polynomials of P.

Stage [1] is the most difficult: we shall come back to its implementation later. Stage [2] is a recursion, and stage [3] is not very difficult. If Q is well chosen, each element of P (seen as a polynomial in x_n) has a constant number of real roots over each component of D', and the surfaces defined by these roots do not intersect. The decomposition of each cylinder induced by these polynomials is therefore cylindrical, and we have found the desired cylindrical decomposition.

It is clear that all the conceptual difficulty in this algorithm arises from the choice of Q. This choice requires a good knowledge of analysis, and of real geometry, which we cannot give here for the general case.

3.2.3.1 *The case of* $\mathbf{R}^2$

For the sake of simplicity, we shall restrict ourselves for the present to the case $n = 2$, writing x and y instead of x_1 and x_2. We begin with the family P, and we have to determine a family Q of polynomials in the variable x, such that a decomposition of $\mathbf{R}^1$ invariant for the signs of Q can serve as the foundation for a cylindrical decomposition of $\mathbf{R}^2$. We may suppose that the elements of P are square-free and relatively prime. What are the restrictions which the components F_j of this decomposition of $\mathbf{R}^1$ have to satisfy?

(1) Over each component, each polynomial of P has a constant number of real roots.

(2) Over each component, the surfaces (in fact, in $\mathbf{R}^2$ they are only curves) given by the roots of the elements of P do not intersect.

For the first restriction, there are two situations in which the number of real (and finite) roots of a polynomial $p_i(x, y)$ (considered as a polynomial in y) could change: a real finite root could become infinite, or it could become complex. For the second situation, we must note that the complex roots of a polynomial with real coefficients always come in pairs, and that therefore *one* real root cannot disappear. There must be a multiple root for a particular value x_0 of x, for two roots of a polynomial equation $p_i(x, y) = 0$ to be able to vanish. The critical values of x which have to appear in the decomposition of $\mathbf{R}^1$ are the values x_0 of x for which one of the polynomials of P has a multiple root. If $p_i(y)$ has a multiple root, $\gcd(p_i, dp_i/dy)$ is not trivial (see the section "Square-free decomposition" in the appendix). In other words, for this value x_0 of x, we have $\mathrm{Res}_y(p_i(x_0, y), dp_i(x_0, y)/dy) = 0$ (see the section on the resultant). But this resultant is a polynomial in x, written $\mathrm{Disc}_y(p_i)$. The critical values are therefore the roots of $\mathrm{Disc}_y(p_i)$ for each element p_i of P. We still have to deal with the possibility that a finite root becomes infinite when the value of x changes. If we write the equation $p(y) = 0$ in the form $a_n(x)y^n + \ldots + a_0(x) = 0$, where the a_i are polynomials in x, and therefore always have finite values, it is obvious that a_n has to cancel to give a root of $p(y)$ which tends to infinity. That can also be deduced from the bounds on the values of roots of polynomials given in the last sub-section.

There are two possibilities for the second restriction also. The two curves which intersect may be roots of the same element or of two different elements of P. In the first case, this element of P has a multiple root where two of its roots intersect; and the case of multiple roots has already been dealt with in the last paragraph. In the second case, the corresponding polynomials must, at the value x_0 of x where these two curves intersect, have a common root, the value of y at the point of intersection. In other words, if p_i and p_j are polynomials, their g.c.d. is not trivial, and $\mathrm{Res}_y(p_i(x_0, y), p_j(x_0, y)) = 0$. The hypothesis that the elements of P have no multiple factors implies that the polynomial $\mathrm{Res}_y(p_i, p_j)$ is not identically zero, and therefore x_0 is determined as a root of the polynomial $\mathrm{Res}_y(p_i, p_j)$.

If we use the notation $\mathrm{lc}(p)$ to denote the leading coefficient of a polynomial p, we have proved the following result.

Proposition. *If $P = \{p_1, \ldots, p_n\}$ is a family of square-free polynomials relatively prime in pairs, it suffices to take for Q the following family:*

$$\{\mathrm{lc}_y(p_i); 1 \leq i \leq n\} \cup \{\mathrm{Disc}_y(p_i); 1 \leq i \leq n\} \cup \{\mathrm{Res}_y(p_i, p_j); 1 \leq i < j \leq n\}.$$

The reader will notice a great increase in the size of the data involved in this reduction to $\mathbf{R}^1$. If the family P consists of m polynomials of degree bounded by d, then Q consists of $O(m^2)$ polynomials of degree bounded by $O(d^2)$. Moreover, the coefficients of the elements of Q will be fairly large — they are bounded by $(d+1)^{2d}B^{2d}$, where B is a bound for the coefficients of the elements of P.

3.2.3.2 *The general case*

The analysis we gave for the case $n = 2$ was rather special to that case. The main principles of the general case are the same, but there are many more possibilities to be considered. The reader should consult the articles of Collins [1975] or of Arnon *et al.* [1984], or the recent results of McCallum [1985a, b]. He has proved that normally* it is enough to take the resultants and the discriminants, as we did in the case of $\mathbf{R}^2$, and also all the coefficients.

Theorem [McCallum, 1985a, b]. *With the same hypotheses as for Collins' theorem, it is possible to calculate a decomposition in time bounded by*

$$(2d)^{n2^{n+7}} m^{2^n} H^3. \tag{6}$$

This result has been slightly improved [Davenport, 1985b], to give

$$(2d)^{n2^{n+5}} m^{2^n} H^3. \tag{7}$$

These growths all behave doubly exponentially with respect to n, and we may ask whether this is really necessary. After all, the problems do not seem to be very complicated. Unfortunately, this behaviour is intrinsic to the problem of calculating a cylindrical decomposition, because of the following result.

Theorem [Davenport and Heintz, 1987]. *There are examples where the space needed to write a cylindrical decomposition, and a fortiori the calculating time, is doubly exponential in n.*

More exactly, if n is of the form $10r+2$, there is a family of $12r$ linear equations and two equations of degree four, such that the family can be

---

* If the primitive part of each polynomial only cancels identically as a polynomial in x_n at a finite number of points of $\mathbf{R}^{n-1}$.

written in $O(n)$ symbols, but such that a cylindrical decomposition of $\mathbf{R}^n$ invariant for its signs needs a space

$$2^{2^{2r+1}} = 2^{2^{(n+3)/5}}. \tag{8}$$

The equations of Davenport and Heintz are specialised, and the result only holds for one choice of co-ordinates, that is of the order of elimination of the variables. In other co-ordinates, the decomposition may be very much simpler. But this proves that the calculation can be very expensive, and the examples in the next section show that this high cost features largely in the applications.

3.2.4 Applications of cylindrical decomposition

There are as many applications as there are polynomial systems with real solutions: we can mention only two of these applications.

3.2.4.1 *Quantifier elimination*

We recall that a quantifier is one of the two symbols $\forall$ (*for all*) and $\exists$ (*there exist*). Quantifiers are used in the form $\forall x \, p(x)$, which means "for all x, the proposition $p(x)$ is true". All the quantified variables in this section are quantified over the real numbers — more formally, we are interested in the first order theory of $\mathbf{R}$. We will use the familiar logical signs $\wedge$ (*and*), $\vee$ (*or*, in the inclusive sense) and $\neg$ (*logical inversion*). This theory contains expressions of the following types, where the x_i are variables, the p_i are polynomials, and A and B are, recursively, expressions of this "theory":

$$p_i(x_i, x_j, \ldots) = 0; \qquad p_i(x_i, x_j, \ldots) > 0;$$
$$A \wedge B; \qquad A \vee B; \qquad \neg A;$$
$$\exists x_i \, A; \qquad \forall x_i \, A.$$

For example, the expression

$$\forall x \, ax^4 + bx^3 + cx^2 + dx + e > 0 \tag{9}$$

is an expression of this theory.

Definition. *The problem of* **quantifier elimination** *is that of finding an algorithm, which, given a logical expression containing quantifiers, finds an equivalent expression which does not contain any.*

We can, for example, ask for an expression equivalent to (9), that is an expression in the variables a, b, c, d and e which is true if and only if the polynomial is always positive.

Tarski [1951] proved that most problems in elementary algebra and in Euclidean geometry can be solved by eliminating quantifiers, and he gave an algorithm for doing it. But his algorithm was completely impractical.

Collins [1975] has used cylindrical decomposition to eliminate quantifiers. The idea is quite simple, although the details are somewhat technical, chiefly because of the possibility of having several quantifiers referring to variables with the same name. So we shall give some examples, rather than a complete algorithm.

Let us consider for example in expression (9) the variables $a,\ldots,e$ as co-ordinates $x_1,\ldots,x_5$ (their order is unimportant); the quantified variable x must be x_6 (here the order is important). Let us make a cylindrical decomposition D of $\mathbf{R}^6$ invariant for the sign of the polynomial of (9). For each component F_i of the cylindrical decomposition of $\mathbf{R}^5$ on which D is based, there is a finite number of components $E_{i,j}$ of D, each with its typical point $z_{i,j}$. If the polynomial of (9) is positive at the typical point $z_{i,j}$, it is positive throughout the component $E_{i,j}$. If the polynomial is positive for all the typical points above F_i, it is positive for all the cylinder above F_i. It follows that the semi-algebraic variety of $\mathbf{R}^5$ which contains exactly the solutions of (9) is

$$\bigcup \{F_i : \forall j a_{i,j} x_{i,j}^4 + b_{i,j} x_{i,j}^3 + c_{i,j} x_{i,j}^2 + d_{i,j} x_{i,j} + e_{i,j} > 0\},$$

where we have written the typical point $z_{i,j}$ as $(a_{i,j}, b_{i,j}, c_{i,j}, d_{i,j}, e_{i,j}, x_{i,j})$. This solution was given by Arnon [1985]. However, none of the above implies that this solution is the simplest solution, nor that this procedure is the best method for finding a solution. This problem was solved with pencil and paper by Mignotte [1986] for the polynomial $x^4 + px^2 + qx + r$, and by Lazard [1987] in the general case (except that he used MACSYMA to expand the necessary resultants). Both found solutions simpler than that obtained algorithmically.

Another problem of this kind is the ellipse problem: what are the conditions on a, b, c and d for the ellipse

$$\left\{(x,y) : \frac{(x-c)^2}{a^2} + \frac{(y-d)^2}{b^2} \leq 1\right\}$$

to be completely contained in the circle $x^2 + y^2 = 1$. Here there are two quantifiers, for the problem can be rewritten in the form

$$\forall x\, \forall y \left(\frac{(x-c)^2}{a^2} + \frac{(y-d)^2}{b^2} \leq 1 \Rightarrow x^2 + y^2 \leq 1\right)$$

This problem has still not been solved in a completely automatic way, although Arnon and Smith [1983] have solved the special case $d = 0$.

Mignotte [1986] has also solved this case with pencil and paper, and Lazard [1987] has solved the general case, using MACSYMA to expand a polynomial T of degree 12 with 105 terms, which features in the solution. The solution he gives has the form

$$T \geq 0 \wedge (c^2 + (b + |d|)^2 \leq 1) \wedge (d^2 + (a + |c|)^2 \leq 1)$$
$$\wedge \Big(\big((b^2 \leq a) \wedge (a^2 \leq b) \big) \vee$$
$$\big((a > b) \wedge (a^2 d^2 \leq (1 - a^2)(a^2 - b^2)) \big) \vee$$
$$\big((b > a) \wedge (b^2 c^2 \leq (1 - b^2)(b^2 - a^2)) \big) \Big),$$

which, as we can see, is not very simple, especially because of the size of T.

We can deduce from these two examples that this tool is very general, but that it seems rather expensive, at least with the existing implementations.

3.2.4.2 *Robotics*

One of the problems of robotics is the problem of *robot motion planning*, that is of deciding how the robot can move from one place to another without touching the walls, without damaging itself on other robots etc. There are, it is true, many other problems, but we shall consider only this one. For the sake of simplicity, we limit ourselves to the case of one robot, and this is difficult enough — the case of several robots is studied in Schwartz and Sharir [1983b]. We suppose that all the obstacles are fixed, and are defined by polynomial equations, and that the robot too is given by polynomial equations.

If we choose a certain number of points on the robot, we can fix exactly the position of each of its parts. For example, if the robot is rigid, it is sufficient to choose three points which are not collinear. If we need k points, the position of the robot is determined by a point in $\mathbf{R}^{3k}$. But there are constraints between these points — for example two points of the same rigid part of the robot are always at the same distance from one another. Thus our points in $\mathbf{R}^{3k}$ have to satisfy a system of polynomial equations (and, possibly, inequalities).

The obstacles define several polynomial inequalities, which have to be satisfied for the position of the robot to be "legitimate", for example, that it does not cross a wall. The laws of physics may introduce several inequalities, for example, the robot must be in a state of equilibrium under the effect of gravity. All these equations and inequalities give us a semi-algebraic variety in $\mathbf{R}^{3k}$. In principle, the problem of motion planning is fairly simple: are the departure position and the desired position connected in the variety? In other words, is there a path in the variety which links the departure position to the desired position? If there is one, then it gives the desired movement; if not the problem has no solution.

This idea has been studied by Schwartz and Sharir [1983a], where they explained how to find the path. Starting from a cylindrical decomposition, the problem breaks down into two stages.

- The geometric aspect: which components are next to one another (we need only consider the components in the variety of legitimate positions).
- The combinatorial aspect: given the graph of the neighbourhood relations between the components, find the paths in this graph which join the component of the departure position and the component of the desired position.

The reader is advised to consult their articles for the details.

Unfortunately, this algorithm scarcely seems practicable. One of the simplest problems of motion planning is the following:

> Given in $\mathbf{R}^2$ a right-angled "corridor" of width unity, say $(x > 0 \wedge 0 \leq y \leq 1) \cup (y > 0 \wedge 0 \leq x \leq 1)$, can we get a rod of length three (and of infinitesimal width) to negotiate it.

The answer is "no", because the biggest rod which can negotiate it is of length $2\sqrt{2}$. Davenport [1986] tried to solve this problem by Schwartz and Sharir's method. The semi-algebraic variety is in $\mathbf{R}^4$ (for we take the two extremities of the rod as points of reference), and is defined by an equation, requiring that these two points be always at a distance three from one another, and by eight inequalities, two for each wall of the corridor. To find a cylindrical decomposition of $\mathbf{R}^4$, we have to determine one of $\mathbf{R}^3$, then one of $\mathbf{R}^2$ and finally one of $\mathbf{R}^1$. For $\mathbf{R}^1$, we find 184 (irreducible) polynomials, of total degree 801, and maximum degree 18. This reduction took ten minutes cpu-time with REDUCE on a DEC 20–60. These polynomials have 375 real roots, which can be isolated (in the sense of isolating the roots of each polynomial) in five minutes. But we have to isolate all the roots of all the polynomials, and Davenport was unable to do the calculation, because of the growth in the numerators and denominators of the rational numbers involved.

4. Advanced algorithms

Computer Algebra, in general, manipulates quite familiar mathematical objects, and the manipulations are, in principle, fairly simple. But as we have already seen, even for calculating the g.c.d. of two polynomials with integer coefficients, naïve algorithms are extremely costly. We therefore introduced methods, such as sub-resultant sequences of polynomials, which seem to be much more efficient. We said:

> This algorithm is the best method known for calculating the g.c.d., of all those based on Euclid's algorithm applied to polynomials with integer coefficients. In the chapter "Advanced algorithms" we shall see that if we go beyond these limits, it is possible to find better algorithms for this calculation.

4.1 MODULAR METHODS

In this section, we shall describe the first (historically speaking) family of non-obvious algorithms, and we shall start with its greatest success: the g.c.d.

4.1.1 g.c.d. in one variable

This section is set in the following context: A and B are two polynomials, belonging to the ring $\mathbf{Z}[x]$, whose g.c.d. we want to calculate. The restriction to integer coefficients does not present any problem, for we can always multiply the polynomials by an integer so as to eliminate any denominator. A slight variation of the classic example of Knuth [1969] (see also Brown [1971]), where this calculation by naïve methods becomes very costly, is:

$$A(x) = x^8 + x^6 - 3x^4 - 3x^3 + x^2 + 2x - 5;$$
$$B(x) = 3x^6 + 5x^4 - 4x^2 - 9x + 21$$

(see the calculations in the section "The g.c.d." of Chapter 2). Let us suppose that these two polynomials have a common factor, that is a polynomial P (of non-zero degree) which divides A and B. Then there is a polynomial Q such that $A = PQ$. This equation still holds if we take each coefficient as an integer modulo 5. If we write P_5 to signify the polynomial P considered as a polynomial with coefficients modulo 5, this equation implies that P_5 divides A_5. Similarly, P_5 divides B_5, and therefore it is a common factor* of A_5 and B_5. But calculating the g.c.d. of A_5 and B_5 is fairly easy:

$$A_5(x) = x^8 + x^6 + 2x^4 + 2x^3 + x^2 + 2x;$$

$$B_5(x) = 3x^6 + x^2 + x + 1;$$

$$C_5(x) = \text{remainder}(A_5(x), B_5(x)) = A_5(x) + 3(x^2 + 1)B_5(x) = 2x^2 + 3;$$

$$D_5(x) = \text{remainder}(B_5(x), C_5(x)) = B_5(x) + (x^4 + x^2 + 3)C_5(x) = x;$$

$$E_5(x) = \text{remainder}(C_5(x), D_5(x)) = C_5(x) + 3xD_5(x) = 3.$$

Thus A_5 and B_5 are relatively prime, which implies that $P_5 = 1$. As the leading coefficient of P has to be one, we deduce that $P = 1$.

The concept of modular methods is inspired by this calculation, where there is no possibility of intermediate expression swell, for the integers modulo 5 are bounded (by 4). Obviously, there is no need to use the integers modulo 5: any prime number p will suffice (we chose 5 because the calculation does not work modulo 2 and 3, for reasons we shall explain later). In this example, the result was that the polynomials are relatively prime. This raises several questions about generalising this calculation to an algorithm capable of calculating the g.c.d. of any pair of polynomials:

(1) how do we calculate a non-trivial g.c.d.?

(2) what do we do if the modular g.c.d. is not the modular image of the g.c.d. (as in the example in the footnote*)?

(3) how much does this method cost?

Before we can answer these questions, we have to be able to bound the coefficients of the g.c.d. of two polynomials.

Theorem (Landau-Mignotte inequality). *Let* $Q = \sum_{i=0}^{q} b_i x^i$ *be a divisor of the polynomial* $P = \sum_{i=0}^{p} a_i x^i$ *(where* a_i *and* b_i *are integers). Then*

$$\sum_{i=0}^{q} |b_i| \leq 2^q \left| \frac{a_p}{b_q} \right| \sqrt{\sum_{i=0}^{p} a_i^2}.$$

See the paper by Landau [1905], and those by Mignotte [1974, 1982].

---

* Note that we cannot deduce that $P_5 = \gcd(A_5, B_5)$: a counter-example is $A = x - 3$, $B = x + 2$, where $P = 1$, but $A_5 = B_5 = x + 2$, and so $\gcd(A_5, B_5) = x - 2$, whereas $P_5 = 1$.

Corollary 1. *Every coefficient of the g.c.d. of $A = \sum_{i=0}^{\alpha} a_i x^i$ and $B = \sum_{i=0}^{\beta} b_i x^i$ (with a_i and b_i integers) is bounded by*

$$2^{\min(\alpha,\beta)} \gcd(a_\alpha, b_\beta) \min \left(\frac{1}{|a_\alpha|} \sqrt{\sum_{i=0}^{\alpha} a_i^2}, \frac{1}{|b_\beta|} \sqrt{\sum_{i=0}^{\beta} b_i^2} \right).$$

Proof. The g.c.d. is a factor of A and of B, the degree of which is, at most, the minimum of the degrees of the two polynomials. Moreover, the leading coefficient of the g.c.d. has to divide the two leading coefficients of A and B, and therefore has to divide their g.c.d.

A slight variation of this corollary is provided by the following result.

Corollary 2. *Every coefficient of the g.c.d. of $A = \sum_{i=0}^{\alpha} a_i x^i$ and $B = \sum_{i=0}^{\beta} b_i x^i$ (where a_i b_i are integers) is bounded by*

$$2^{\min(\alpha,\beta)} \gcd(a_0, b_0) \min \left(\frac{1}{|a_0|} \sqrt{\sum_{i=0}^{\alpha} a_i^2}, \frac{1}{|b_0|} \sqrt{\sum_{i=0}^{\beta} b_i^2} \right).$$

Proof. If $C = \sum_{i=0}^{\gamma} c_i x^i$ is a divisor of A, then $\hat{C} = \sum_{i=0}^{\gamma} c_{\gamma-i} x^i$ is a divisor of $\hat{A} = \sum_{i=0}^{\alpha} a_{\alpha-i} x^i$, and conversely. Therefore, the last corollary can be applied to $\hat{A}$ and $\hat{B}$, and this yields the bound stated.

It may seem strange that the coefficients of a g.c.d. of two polynomials can be greater than the coefficients of the polynomials themselves. One example which shows this is the following (due to Davenport and Trager):

$$A = x^3 + x^2 - x - 1 = (x+1)^2(x-1);$$
$$B = x^4 + x^3 + x + 1 = (x+1)^2(x^2 - x + 1);$$
$$\gcd(A, B) = x^2 + 2x + 1 \qquad = (x+1)^2.$$

This example can be generalised, as say

$$A = x^5 + 3x^4 + 2x^3 - 2x^2 - 3x - 1 \qquad = (x+1)^4(x-1);$$
$$B = x^6 + 3x^5 + 3x^4 + 2x^3 + 3x^2 + 3x + 1 = (x+1)^4(x^2 - x + 1);$$
$$\gcd(A, B) = x^4 + 4x^3 + 6x^2 + 4x + 1 \qquad = (x+1)^4.$$

4.1.1.1 *The modular integer relationship*

In this sub-section, we answer the question raised above: what do we do if the modular g.c.d. is not the modular image of the g.c.d. calculated over the integers?

Lemma 1. *If p does not divide the leading coefficient of* $\gcd(A, B)$, *the degree of* $\gcd(A_p, B_p)$ *is greater than or equal to that of* $\gcd(A, B)$.

Proof. Since $\gcd(A, B)$ divides A, then $(\gcd(A, B))_p$ divides A_p. Similarly, it divides B_p, and therefore it divides $\gcd(A_p, B_p)$. This implies that the degree of $\gcd(A_p, B_p)$ is greater than or equal to that of $\gcd(A, B)_p$. But the degree of $\gcd(A, B)_p$ is equal to that of $\gcd(A, B)$, for the leading coefficient of $\gcd(A, B)$ does not cancel when it is reduced modulo p.

This lemma is not very easy to use on its own, for it supposes that we know the g.c.d. (or at least its leading coefficient) before we are able to check whether the modular reduction has the same degree. But this leading coefficient has to divide the two leading coefficients of A and B, and this gives a formulation which is easier to use.

Corollary. *If p does not divide the leading coefficients of A and of B (it may divide one, but not both), then the degree of* $\gcd(A_p, B_p)$ *is greater than or equal to that of* $\gcd(A, B)$.

As the g.c.d. is the only polynomial (to within an integer multiple) of its degree which divides A and B, we can test the correctness of our calculatio–s of the g.c.d.: if the result has the degree of $\gcd(A_p, B_p)$ (where p satisfies the hypothesis of this corollary) and if it divides A and B, then it is the g.c.d. (to within an integer multiple).

It is quite possible that we could find a $\gcd(A_p, B_p)$ of too high a degree. For example, in the case we have already cited of

$$A(x) = x^8 + x^6 - 3x^4 - 3x^3 + x^2 + 2x - 5;$$
$$B(x) = 3x^6 + 5x^4 - 4x^2 - 9x + 21,$$

$\gcd(A_2, B_2) = x + 1$ (it is obvious that $x + 1$ divides the two polynomials modulo 2, because the sum of the coefficients of each polynomial is even). The following lemma shows that this possibility can only arise for a finite number of p.

Lemma 2. *Let* $C = \gcd(A, B)$. *If p satisfies the condition of the corollary, and if p does not divide* $\mathrm{Res}_x(A/C, B/C)$, *then* $\gcd(A_p, B_p) = C_p$.

Proof. A/C and B/C are relatively prime, for otherwise C would not be the g.c.d. of A and B. By the corollary, C_p does not vanish. Therefore

$$\gcd(A_p, B_p) = C_p \gcd(A_p/C_p, B_p/C_p).$$

For the lemma to be false, the last g.c.d. has to be non-trivial. This implies that the resultant $\mathrm{Res}_x(A_p/C_p, B_p/C_p)$ vanishes, by proposition 1 of the section "The resultant" in the appendix "Algebraic background". This resultant is the determinant of a Sylvester matrix, and $|M_p| = (|M|)_p$, for the determinant is only a sum of products of the coefficients. In the present case, this amounts to saying that $\text{Res}_x(A/C, B/C)_p$ vanishes, that is that p divides $\text{Res}_x(A/C, B/C)$. But the hypotheses of the lemma exclude this possibility.

Definition. *If* $\gcd(A_p, B_p) = \gcd(A, B)_p$, *we say that the reduction of this problem is good, or that* p *is of good reduction. If not, we say that* p *is of bad reduction.*

This lemma implies, in particular, that there are only a finite number of values of p such that $\gcd(A_p, B_p)$ does not have the same degree as that of $\gcd(A, B)$, that is the p which divide the g.c.d. of the leading coefficients and the p which divide the resultant of the lemma (the resultant is non-zero, and therefore has only a finite number of divisors). In particular, if A and B are relatively prime, we can always find a p such that A_p and B_p are relatively prime.

4.1.1.2 *Calculation of the g.c.d.*

In this section we answer the question posed earlier: how do we calculate a non-trivial g.c.d.? One obvious method is to use the Landau-Mignotte inequality, which can determine an M such that all the coefficients of the g.c.d. are bounded by M, and to calculate modulo a prime number greater than $2M$. This method translates into the following algorithm:

$M := Landau\_Mignotte\_bound(A, B);$
$\textbf{do } p := find\_large\_prime(2M);$
$\quad \textbf{if } degree\_remainder(p, A) \textbf{ or } degree\_remainder(p, A)$
$\quad\quad \textbf{then } C := modular\_gcd(A, B, p);$
$\quad\quad\quad \textbf{if } divide(C, A) \textbf{ and } divide(C, B)$
$\quad\quad\quad\quad \textbf{then return } C;$
$\quad \textbf{forever};$

(where the algorithm *Landau_Mignotte_bound* applies the corollaries of their inequality, the algorithm *find_large_prime* returns a prime number greater than its argument (a different number each time), the algorithm *degree_remainder* verifies that the reduction modulo p does not change the degree, that is that p does not divide the leading coefficient, the algorithm *modular_gcd* applies Euclid's algorithm modulo p and the algorithm *divide* verifies that the polynomials divide over the integers). In fact, it is not necessary to test that p does not divide the leading coefficients — the Landau-Mignotte bound (corollary 1) implies that p is greater than one of the leading coefficients of A and B.

The drawback to this method is that it requires lengthy calculations modulo p, which may be a fairly large integer. So we suggest a method which uses several small primes and the Chinese remainder theorem (see the appendix "Algebraic background"). If we calculate C_p and C_q, where C is the desired g.c.d., and p and q are two primes, then this theorem calculates C_{pq} for us. We must point out that this theorem is applied in its integer form, to each coefficient of C separately. There is no question of using the polynomial version, even though we are trying to recover a polynomial. This method translates into the following algorithm:

$M := Landau\_Mignotte\_bound(A, B);$
$Avoid := \gcd(\mathrm{lc}(A), \mathrm{lc}(B));$
E0: $p := find\_prime(Avoid);$
$C := modular\_gcd(A, B, p);$
E1: **if** $degree(C) = 0$ **then return** 1;
$Known := p;$
$Result := C;$
while $Known \leq 2M$
$\qquad$ **do** $p := find\_prime(Avoid);$
$\qquad\quad$ $C := modular\_gcd(A, B, p);$
$\qquad\quad$ **if** $degree(C) < degree(Result)$ **then go to** *E1*;
$\qquad\quad$ **if** $degree(C) = degree(Result)$
$\qquad\qquad$ **then** $Result := CRT(Result, Known, C, p);$
$\qquad\qquad\quad$ $Known := Known \times p;$
$\quad$ **if** $divide(Result, A)$ **and** $divide(Result, B)$
$\qquad$ **then return** $Result;$
$\quad$ **go to** *E0;*

(where "lc" denotes the leading coefficient of a polynomial, the sub-algorithms of the last algorithm have the same meaning here, and the algorithm *find_prime* returns a prime which does not divide its argument (a different prime each time), and the algorithm *CRT* applies the Chinese remainder theorem to each coefficient of the the two polynomials *Result* (modulo *Known*) and C (modulo p), representing the integers modulo M between $-M/2$ and $M/2$). The two **go to*** in this algorithm correspond to the two ways of detecting that all the chosen primes were of bad reduction:

- either we find a p (satisfying the corollary of lemma 1) such that the degree of the g.c.d. modulo p is smaller than the degrees already calculated (and therefore they come from bad reductions);

---

* which could be replaced by loops, or other tools of structured programming, if one desired.

- or we get to the end of the calculations with a result which seems good, but which does not divide A and B, since all the reductions have been bad (an unlikely possibility).

If the first reduction was good, no **go to** would be executed.

This algorithm is open to several improvements. It is not necessary for the p to be genuinely primes. It suffices if the infinite set of the p has an infinite number of prime factors, for we know that there is only a finite number of values with bad reduction. If p is not a prime, it is possible that the algorithm *modular-gcd* finds a number which is not zero, but which cannot be inverted modulo p, such as 2 (mod 6). This discovery leads to a factorisation of the modulus, and we can therefore work modulo the two factors separately. The line

$$Known := Known \times p;$$

should of course be replaced by

$$Known := \mathrm{lcm}(Known, p);$$

and the implementation of the Chinese remainder theorem should be capable of dealing with moduli which are not relatively prime.

It is not absolutely necessary to go as far as $Known > 2M$. The Landau-Mignotte bounds are often far too pessimistic, although Mignotte [1981] proved that they cannot, in general, be reduced. It is possible to test at each stage whether we have found a common divisor, for we know that this method can never find a polynomial of too small a degree. But these tests can be very costly, and a compromise often used is to test after those stages in which the value does not change, that is to replace the line

$$\textbf{then } Result := CRT(Result, Known, C, p);$$

by the lines

$$\textbf{then } Previous := Result;$$
$$Result := CRT(Result, Known, C, p);$$
$$\textbf{if } Previous = Result \textbf{ and}$$
$$divide(Result, A) \textbf{ and } divide(Result, B)$$
$$\textbf{then return } Result;$$

4.1.1.3 *Cost of this algorithm*

We now analyse the basic algorithm, as we have written it, without the improvements outlined at the end of the last paragraph. We add the following hypotheses.

(1) All the primes have already been calculated (this hypothesis is not very realistic, but we have outlined a method which can use numbers which are not prime, and in practice this search for primes is not very expensive);

(2) Each prime p is sufficiently small for all the calculations modulo p to be done in time $O(1)$; that is, the numbers modulo p can be dealt with directly on the computer, for one word is enough to store them;

(3) The polynomials A and B are of degree less than or equal to n, and satisfy the bound

$$\sqrt{\sum_{i=0}^{\alpha} a_i^2}, \sqrt{\sum_{i=0}^{\beta} b_i^2} \le H$$

(which would be satisfied if all the coefficients were less than $H/\sqrt{n+1}$, and which implies that all the coefficients are less than H).

The coefficients of the g.c.d. are bounded by $2^n H$, according to the Landau-Mignotte inequality. If N_1 is the number of p with good reduction we need, the product of all these p must be greater than $2^{n+1} H$, and this implies that $N_1 < (n+1) \log_2 H$ (where we use the inequality that every prime is at least 2: a more delicate analysis would bring in analytic number theory without improving the order obtained for the running time). Moreover, there are the p of bad reduction — let us suppose that there are N_2 of them. They must divide the resultant of A/C and B/C. These two polynomials are factors of A and B, and therefore their coefficients (in fact, the sum of the absolute values of the coefficients) are bounded by $2^n H$. Their resultant is, therefore, bounded by $(2^n H)^{2n}$. This implies $N_2 < n(n + \log_2 H)$. We can deduce a better bound $N_2 < n \log_2 H$, if we note that "bad reduction" is equivalent to saying that one of the minors of Sylvester's matrix is divisible by p, even though it is not zero (see Loos [1982]).

The big **while** loop is therefore executed at most $N_1 + N_2$ times — probably fewer, for we have supposed that the resultant had all its factors distinct and small, and that we meet all its factors before finding enough p of good reduction. The expensive operations of this loop are the modular g.c.d. and the application of the Chinese remainder theorem. For the modular g.c.d., we first have to reduce the coefficients of A and B modulo p — $O(n \log_2 H)$ operations — and then to calculate the g.c.d. — $O(n^2)$ operations. The Chinese remainder theorem is applied to all the coefficients of the supposed g.c.d. We write C for the g.c.d. modulo p, and D for the g.c.d. modulo M, the product of all the p already calculated (written *Known* in the algorithm). To prove the theorem we calculate integers f and g such that $fM + gp = 1$, and prove that $e = c + (d - c)fM$ is the value modulo

Mp which reduces to c modulo M and d modulo p. Calculating f and* g requires $O(\log_2 M)$ operations (for p is supposed small, by hypothesis 2). f is bounded by p, and $b - a$ can be calculated modulo p (for the result is modulo MP). The multiplication $(b-a)fM$ is the multiplication of a large number M by a small number, and requires $O(\log_2 M)$ operations. As we have at most $n + 1$ coefficients to calculate, the total cost of a cycle of the loop is $O(n(n + \log_2 H) + n \log_2 M)$ operations. As $M < 2^{n+1}H$, this can be simplified to $O(N(n + \log_2 H))$.

Thus we deduce a total cost of $O(n(n + \log_2 H)^2)$ operations for the loop. The last stage of this algorithm is checking the result, by dividing A and B by the result, and this checks that we have found a common divisor of A and B. The most expensive case is the division of a polynomial of degree n by a polynomial of degree $n/2$, and this requires $n^2/4$ operations by naïve methods. As these operations were performed on integers of size (number of digits) bounded** by $n \log_2 H$, we find that the total cost of these divisions is $O(n^4 \log_2^2 H)$ operations. This is somewhat annoying, since this implies that the verification is considerably more costly than the calculation itself. We can use "efficient" methods (based on the Fast Fourier Transform), and this gives us

$$O\left(n^3 \log_2 H \log_2 n \log_2(n \log_2 H) \log_2 \log_2(n \log_2 H)\right)$$

operations, but this notation conceals a fairly large constant, and the cost remains higher than that for the loop. Brown [1971] therefore proposed continuing the loop until we are sure of having exhausted all the p of bad reduction, that is adding to the condition of the **while** the phrase

or $\prod p < Resultant\_Bound$

where the product extends over all the p already provided by *find_prime*, and the variable *Resultant_Bound* is given the value of a bound for the

---

* Although we do not use g in this calculation, the extended Euclidean algorithm calculates it automatically. It need only be calculated once, rather than once per coefficient. This does not alter the asymptotic cost, but it is quite important in practice, for the constant concealed by the notation O is considerably larger for the extended Euclidean algorithm than that for multiplication.

** If we find an integer of the quotient greater than this bound, this implies that the division is not exact, and therefore we can terminate the operation *divide* immediately. This remark is quite important in practice, for a division of this kind which fails may generate huge coefficients — for example the division of x^{100} by $x - 10$ gives a remainder 10^{100}. See Abbott *et al.* [1985] for a discussion of the cost of the operation *divide*.

resultant of A and B, and this limits also all the minors, as the cancellation of one of them was the condition of bad reduction. With this addition, we know that we have chosen a p of good reduction, and that the final result is correct. This addition is of theoretical use, reducing the calculating time to $O(n^3 log_2^3 H)$, but implementers prefer the algorithm in the form in which we stated it.

In practice, we can suppose that the algorithm does not find many p of bad reduction. If, in addition, we suppose that the coefficients of the g.c.d. are less than H, and if we add a termination test (such as that outlined in the preceding paragraph), we arrive at a cost of $O(n \log_2 H (n + \log_2 H))$. In any case this algorithm is asymptotically more efficient than the sub-resultant algorithm ($O(n^4 \log_2^4 H)$ according to Loos [1982]). This efficiency is borne out in practice.

4.1.2 g.c.d. in several variables

The previous section showed the use of modular calculations for avoiding intermediate expression swell. When we calculate modulo p, we have a guarantee that the integers do not exceed $p - 1$ (or $p/2$ if we are using a symmetric representation). This method gives an algorithm for calculating g.c.d.s in one variable which is more efficient than the sub-resultant polynomial sequences algorithm. Can it be generalised to polynomials in several variables? Before we can answer this question, we must settle some details.

The most obvious algorithm for calculating the g.c.d. in several variables $x_1, \ldots, x_n$ is to convert Euclid's algorithm into $\mathbf{Q}(x_1, \ldots, x_{n-1})[x_n]$. But this has one big theoretical drawback (and several practical ones): it calculates only the dependence of the g.c.d. on x_n. For example, if it is applied to

$$A(x_1, x_2) = (x_1 - 1)x_2 + (x_1 - 1),$$
$$B(x_1, x_2) = (x_1 - 1)x_2 + (-x_1 + 1),$$

the answer would be that they are relatively prime, even though they have a factor of $x_1 - 1$ in common.

Definition. *Let R be an integral domain, and $p \in R[x]$, with coefficients $a_0, \ldots, a_k$, such that $p = \sum_{i=0}^{k} a_i x^i$. The content of p, written $\mathrm{cont}(p)$, is the g.c.d. of all its coefficients. If the content of a polynomial is one, the polynomial is called primitive. The primitive part of a polynomial p, written $\mathrm{pp}(p)$, is defined by $\mathrm{pp}(p) = p/\mathrm{cont}(p)$.*

It is easy to deduce that the primitive part of a polynomial is primitive. The following result is quite well known in mathematical literature, and it demonstrates the great importance of this idea. In the section on factorisation, we shall present this lemma in a slightly different form.

Gauss's Lemma. *Let p and q be two polynomials of $R[x]$. Then* $\mathrm{cont}(pq)$ $= \mathrm{cont}(p)\,\mathrm{cont}(q)$ *(and therefore* $\mathrm{pp}(pq) = \mathrm{pp}(p)\,\mathrm{pp}(q)$*).*

Corollary. *Let p and q be two polynomials of $R[x]$. Then*

$$\mathrm{cont}\gcd(p,q) = \gcd(\mathrm{cont}(p),\mathrm{cont}(q)),$$

$$\mathrm{pp}\gcd(p,q) = \gcd(\mathrm{pp}(p),\mathrm{pp}(q)).$$

Thus, given two polynomials in several variables, we can call one variable the "main variable" (written x), and we can calculate the g.c.d. by multiplying the g.c.d. of the primitive parts (found by Euclid's algorithm, taking care that the result is primitive) by the g.c.d. of the contents. This second g.c.d., as well as the g.c.d.s needed to calculate the contents, do not involve the variable x, and can therefore be done recursively.

The context of this section is as follows: A and B are two polynomials belonging to the ring $\mathbf{Z}[x_1,\ldots,x_r]$ whose g.c.d. we want to calculate. Restricting ourselves to polynomials and to integer coefficients does not cause any problems, for we can always multiply the polynomials by an integer to eliminate a numerical denominator, and by a polynomial in several x_i to eliminate a polynomial denominator. Therefore, the procedure in the last paragraph can be expressed in terms of two recursive algorithms:

```
proc gcd(A, B, r);
     Ac; = content(A, r);
     Ap; = A/Ac;
     Bc; = content(B, r);
     Bp; = B/Bc;
     return content(Euclid(Ap, Bp, r), xr) × gcd(Ac, Bc, r − 1);

proc content(A, r);
     Result := coeff(A, xr, 0);
     i := 1;
     while Result ≠ 1 and i < degree(A, xr)
          do Result := gcd(Result, coeff(A, xr, 0), r − 1);
             i := i + 1;
     return Result;
```

where the operators *degree* and *coeff* extract the indicated components from their first parameter, considered as a polynomial in the second, and the algorithm *Euclid* applies this algorithm to its first two parameters, considered as polynomials in one variable, that is the third parameter. Here we have chosen to suppress the content of A and B before applying Euclid's algorithm — this choice does not alter the result, but it makes the *Euclid* parameters smaller. The algorithm *content* stops as soon as it finds that the content is one: this can save useless calculations. The problems with this algorithm are many.

(a) The intermediate expression swell in Euclid's algorithm. Even if we use the method of sub-resultant polynomial remainder sequences, the intermediate degrees will grow considerably.

(b) The number of recursive calls. If we want to calculate the g.c.d. of two polynomials in two variables, of degree ten in each one, so that the g.c.d. is of degree five in each variable, we need:

   20 calls to calculate the two contents;

   1 call to calculate the g.c.d. of the two contents;

   5 calls to calculate the content of the result of *Euclid*;

   that is 26 recursive calls. The case of the three variables therefore needs 26 calls on the algorithm *gcd* in two variables, each needing about 26 calls on *gcd* in one variable.

(c) Moreover, all the integers appearing as coefficients can become very large.

These drawbacks are not purely hypothetical. To illustrate the enormous swell which occurs we give a very small example, chosen so that the results will be small enough to be printed, in two variables x and y (x being the main variable):

$$A = (y^2 - y - 1)x^2 - (y^2 - 2)x + (y^2 + y + 1);$$
$$B = (y^2 - y + 1)x^2 - (y^2 + 2)x + (y^2 + y + 2).$$

The first elimination gives

$$C = (2y^2 + 4y)x + (2y^4 + y^2 - 3y + 1),$$

and the second gives

$$D = -4y^{10} + 4y^9 - 4y^8 + 8y^7 - 7y^6 + y^5 - 4y^4 + 7y^3 - 14y^2 + 3y - 1.$$

Here we see a growth in the size of the integers, and a very significant growth in the power of y. Note that a generic problem of this degree can result in a polynomial of degree 32 in y.

As we saw in the last section, modular calculation can avoid this growth of integer coefficients. But we have to be more subtle to get round the other problems of growth of the powers of non-principal variables, and of recursive calls. Let us suppose that the two polynomials A and B of our example have a common factor, that is a polynomial P (of non-zero degree) which divides A and B. Then there exists a polynomial Q such that $A = PQ$. This equation is still true if we evaluate every polynomial to the value $y = 2$. If we write $P_{y=2}$ to signify the polynomial P evaluated to the value $y = 2$, this equation implies that $P_{y=2}$ divides $A_{y=2}$. Similarly, $P_{y=2}$ divides $B_{y=2}$, and is therefore a common factor* of $A_{y=2}$ and $B_{y=2}$. But the calculation of the g.c.d. of $A_{y=2}$ and $B_{y=2}$ is quite simple:

$$A_{y=2}(x) = x^2 - 2x + 7;$$
$$B_{y=2}(x) = 3x^2 - 6x + 8;$$
$$\text{remainder}(A_{y=2}(x), B_{y=2}(x)) = 29.$$

Thus $A_{y=2}$ and $B_{y=2}$ are relatively prime, and this implies that $P_{y=2} = 1$. Since the leading coefficient of P has to divide the g.c.d. of the leading coefficients of A and B, which is one, we deduce that $P = 1$.

This calculation is quite like the calculation modulo 5 which we did in the last section, and can be generalised in the same way. The reader who is interested in seeing mathematics in the most general setting should consult the article by Lauer [1982], which gives a general formalisation of modular calculation.

4.1.2.1 *Bad reduction*

We have defined a number p as being *of bad reduction* for the calculation of the g.c.d. of A and B if

$$\gcd(A, B)_p \neq \gcd(A_p, B_p).$$

We have also remarked that, if p is not of bad reduction, and if one of the leading coefficients of A and B is not divisible by p, then the g.c.d. modulo p has the same degree as the true g.c.d. These remarks lead to the following definition.

Definition. *Let A and B be two polynomials of $R[y][x]$, where R is a integral domain. The element r of R is of good reduction for the calculation of the g.c.d. of A and B if*

$$\gcd(A, B)_{y=r} = \gcd(A_{y=r}, B_{y=r}).$$

Otherwise, r is of bad reduction.

Proposition. *r is of bad reduction if and only if $y - r$ divides*

$$\text{Res}_x \left(\frac{A}{\gcd(A, B)}, \frac{B}{\gcd(A, B)} \right).$$

---

* As in the previous section, it must be observed that we cannot deduce that $P_{y=2} = \gcd(A_{y=2}, B_{y=2})$: a counter example is $A = yx + 2$, $B = 2x + y$, where $P = 1$, but $A_{y=2} = B_{y=2} = 2x + 2$, and thus $\gcd(A_{y=2}, B_{y=2}) = x + 1$, whereas $P_{y=2} = 1$.

Proposition. *If r is of good reduction and $y - r$ does not divide the two leading coefficients (with respect to x) of A and B, $\gcd(A, B)$ and $\gcd(A_{y=r}, B_{y=r})$ have the same degree with respect to x.*

The proofs of these propositions are similar to those of the previous section. A key result of that section was the Landau-Mignotte inequality, which is rather difficult to prove, and rather surprising, for the obvious conjecture is false. In the present case, things are very much simpler.

Proposition. *If C is a factor of A, then the degree (with respect to y) of C is less than or equal to that of A.*

4.1.2.2 *The algorithm*

Armed with these results, we can calculate the g.c.d. of two polynomials in n variables by a recursive method, where r is the index of the leading variable (which will not change) and s is the index of the second variable, which will be evaluated. The algorithm is given on the next page. The initial call has to be done with $s = r - 1$. When s drops to zero, our polynomial is in one variable, the recursion is terminated, and we call on a *gcd_simple* procedure (which may well be the modular g.c.d. procedure of the last section) to do this calculation.

In this algorithm, we have supposed that the *random* algorithm returns a random integer (a different number each time), and the *CRT* algorithm applies the Chinese remainder theorem (polynomial version) to each coefficient (with respect to x_r) of the two polynomials *Res* (modulo *Known*) and C (modulo $x_s - v$). The two loops **do ... while**, which are identical, choose a value v which satisfies the condition that $x_s - v$ does not divide the two leading coefficients of A and B.

As in the modular g.c.d. algorithm of the last section, the two **go to** in this algorithm correspond to two ways of detecting whether all the chosen values were of bad reduction:

- either we find a v such that the degree of the g.c.d. after the evaluation $x_s = v$ is smaller than the degrees already calculated (and that they therefore come from bad reductions);
- or else we get to the end of the calculation with a result which looks good but which does not divide A and B because all the reductions have been bad (a rather unlikely possibility).

If the first reduction was good, no **go to** would be carried out.

This algorithm can be improved upon in the same way as the algorithm for the case of one variable, by replacing the line

$$\textbf{then } Result := CRT(Result, Known, C, x_s - v);$$

Algorithm of modular g.c.d.

---

   proc $gcd(A, B, r, s)$;
    if $s = 0$ **then return** $gcd\_simple(A, B, x_r)$;
    $M := 1 + \min(degree(A, x_s), degree(B, x_s))$;
E0:   **do** $v := random()$;
     $Av := subs(x_s, v, A)$;
     $Bv := subs(x_s, v, B)$;
    while $degree(Av, x_r) \neq degree(A, x_r)$
      and $degree(Bv, x_r) \neq degree(B, x_r)$
    $C := gcd(A, B, r, s - 1)$;
E1: $Known := x_s - v$;
    $n := 1$;
    $Res := C$;
    while $n \leq M$
      do do $v := random()$;
       $Av := subs(x_s, v, A)$;
       $Bv := subs(x_s, v, B)$;
      while $degree(Av, x_r) \neq degree(A, x_r)$
        and $degree(Bv, x_r) \neq degree(B, x_r)$
     $C := gcd(A, B, r, s - 1)$;
     if $degree(C, x_r) < degree(Res, x_r)$ **then go to** *E1*;
     if $degree(C, x_r) = degree(Res, x_r)$
      then $Res := CRT(Res, Known, C, x_s - v)$;
       $Known := Known \times (x_s - v)$;
   if $divide(Res, A)$ **and** $divide(Res, B)$
    then return Res;
   go to *E0;*

---

by the lines

    then *Previous* $:= Result$;
     $Result := CRT(Result, Known, C, x_s - v)$;
     if *Previous* $= Result$ **and**
      $divide(Result, A)$ **and** $divide(Result, B)$
     then return *Result*;

4.1.2.3 *Cost of this algorithm*

We now analyse the naïve algorithm as it is written, without the improvement described at the end of the last paragraph. We use the following notation: r means the total number of variables (thus, after $r - 1$ recursions, we shall call the algorithm *gcd_simple*), the degrees with respect to each variable are bounded by n, and the lengths of all the coefficients are bounded by d.

We need at most $n + 1$ values of good reduction. We can leave out of account the small loops **do** ... **while**, for in total, they repeat at most n times, for a repetition means that the value of v returned by *random* is a common root of the leading coefficients of A and B, whereas a polynomial of degree n cannot have more than n roots. If N is the number of values of x_s of bad reduction we find, then the big loop is gone through $N + n + 1$ times. It is then necessary to bound N, which is obviously less than or equal to the number of v of bad reduction.

Let C be the true g.c.d. of A and B. Thus, A/C and B/C are relatively prime, and their resultant (see the appendix "Algebraic background") with respect to x_r is non-zero. If v is of bad reduction, $A_{x_s=v}$ and $B_{x_s=v}$ have a g.c.d. greater than $C_{x_s=v}$, i.e. $\left(\frac{A}{C}\right)_{x_s=v}$ and $\left(\frac{B}{C}\right)_{x_s=v}$ are not relatively prime, and therefore their resultant with respect to x_r must be zero. But if the leading coefficients do not cancel,

$$\mathrm{Res}_{x_r}\left(\frac{A}{C}, \frac{B}{C}\right)_{x_s=v} = \mathrm{Res}_{x_r}\left(\left(\frac{A}{C}\right)_{x_s=v}, \left(\frac{B}{C}\right)_{x_s=v}\right).$$

(If one of the leading coefficients (for example, that of A) cancels, the situation is somewhat more complicated. The resultant on the right is the determinant of a Sylvester matrix of dimension smaller than that on the left. In this case, the resultant on the right, multiplied by a power of the other leading coefficient, is equal to the resultant on the left. In any case, one resultant cancels if and only if the other cancels.) This implies that v is a root of $\mathrm{Res}_{x_r}(A/C, B/C)$. This is a polynomial of degree at most $2n^2$, and therefore $N \leq 2N^2$.

In this loop, the most expensive operations are the recursive call to the algorithm *gcd* and the application of the Chinese remainder theorem. For the latter, we have to calculate at most $n + 1$ coefficients of a polynomial in x_r, by applying the theorem to *Known* and $x_s - v$. *Known* is a product of linear polynomials, and the section on the Chinese remainder theorem in the appendix "Algebraic background" gives an extremely simple formulation in this case. In fact, we only have to evaluate *Res* at the value v, to multiply the difference between it and C by a rational number and by *Known*, and to add this product to *Res*. The multiplication (the most expensive operation) requires a running time $O(n^{2s-1}d^2)$ for each coefficient with respect to the calculated x_r, that is a total time of $O(n^{2s}d^2)$. Thus, if we denote by gcd(s) the running time of the algorithm *gcd* with fourth parameter equal to s, the cost of one repetition of the loop is $\mathrm{gcd}(s - 1) + O(n^{2s}d^2)$.

The total cost therefore becomes

$$(2n^2 + n + 1)\left(\mathrm{gcd}(s - 1) + O(n^{2s}d^2)\right).$$

The results of the last section, which dealt with the case of one variable, imply that $\gcd(0) = O(n^3 d^3)$. An induction argument proves that $\gcd(s) = O(n^{2s+3}d^3)$. The initial value of s is $r-1$, and therefore our running time is bounded by $O(n^{2r+1}d^3)$. This is rather pessimistic, and Loos [1982] cites a conjecture of Collins that on average the time is rather $O(n^{r+1}d^2)$.

4.1.3 Other applications of modular methods

Modular methods have many other applications than g.c.d.s, although this was the first one. Some of these applications also have the idea of bad reduction, for others the reductions are always good. Obviously, the algorithm is much simpler if the reductions are always good.

In general, we illustrate these applications by reduction modulo p, where p is a prime number, as we did in the first section. If the problems involve several variables, we can evaluate the variables at certain values, as we did for the g.c.d. in several variables. The generalisation is usually fairly obvious, and can be looked up in the references cited.

4.1.3.1 *Resultant calculation*

Resultants are defined in the appendix "Algebraic background". As this calculation is very closely linked to that of the g.c.d., we should not be surprised by the existence of a link between the two. In fact, as the resultant of two polynomials A and B is a polynomial in the coefficients of A and B, we see (at least if p does not divide the leading coefficients) that

$$\operatorname{Res}_x(A, B)_p = \operatorname{Res}_x(A_p, B_p).$$

As this equation always holds, every p is of good reduction. We exclude those p which divide the leading coefficients in order to avoid the complications that arise when the degrees of A and B change. The fore-mentioned Appendix gives bounds on the size of a resultant, and they tell us how many p are needed to guarantee that we have found the resultant. Collins [1971] discusses this algorithm, and it turns out to be the most efficient method known for this calculation.

4.1.3.2 *Calculating determinants*

The resultant is only the determinant of the Sylvester matrix, and we may hope that the modular methods will let us calculate general determinants. This is in fact possible — all we have to do is to calculate the determinant modulo a sufficient number of p, and apply the Chinese remainder theorem. We have to determine the number of p required — in other words, we must bound the determinant. Let us suppose that the determinant is of dimension n, and that the entries are $a_{i,j}$.

Theorem.

$$|\det(a_{i,j})| \leq n! \left(\max_i \max_j |a_{i,j}| \right)^n .$$

Proof. The determinant consists of the sum of $n!$ terms, of which each is the product of n elements of the matrix.

Theorem (Hadamard's bound [1893]).

$$|\det(a_{i,j})|^2 \leq \prod_{i=1}^{n} \left(\sum_{j=1}^{n} |a_{i,j}|^2 \right).$$

4.1.3.3 *Inverse of a matrix*

This problem is rather more complicated, for the elements of the inverse of a matrix with integer coefficients are not necessarily integers. Moreover, it is possible that the matrix can be inverted over the integers, but that it is singular modulo p, and therefore bad reduction can occur. If inversion does not work modulo p, this implies that p divides the determinant, and, if the matrix is not truly singular, there are only finitely many such p. Therefore, if we avoid such p, inversion is possible. There are two solutions to the problem set by the fact that the elements of the inverse are not necessarily integers. The first is to note that they are always rational numbers whose denominators are divisible by the determinant. Therefore, if we multiply the element modulo p by the determinant (which we calculate once and for all) modulo p, we get back to the numerator modulo p, which is used to find the true solution.

The other solution is important, because it can be applied to other problems where the desired solutions are fractions. Wang [1981] suggested this method in the context of the search for decompositions into partial fractions, but it can be applied more generally. First we need some algebraic background. Let M be an integer (in practice, M will be the product of all the primes of good reduction we have used), and a/b a rational number such that b and M are relatively prime (otherwise, one of the reductions was not good). If we apply the extended Euclidean algorithm (see the appendix "Algebraic background") to b and M, we find integers c and d such that $bc + Md = 1$ — in other words $bc \equiv 1 \pmod{M}$. We can extend the idea to a rational congruence by saying

$$\frac{a}{b} \equiv ac \pmod{M}.$$

Moreover, if a and b are less than $\sqrt{M/2}$, this representation is unique, for $a/b \equiv a'/b'$ implies $ab' \equiv a'b$. For example, if $M = 9$, we can represent all the fractions with numerator and denominator less than or equal to 2, as we see in the following table:

Modulo 9	Fraction
0	$0/0$
1	$1/1 = 2/2$
2	$2/1$
3	—
4	$-1/2$
5	$1/2$
6	—
7	$-2/1$
8	$-1/1 = -2/2$

We have proved the second clause of the following principle.

Principle of Modular Calculation. *An integer less than $\frac{1}{2}M$ has a unique image n modulo M. Similarly, a rational number whose numerator and denominator are less than $\sqrt{M/2}$ has a unique image modulo M.*

In the case of integers, given the image n, calculating an integer is fairly simple — we take n or $n - M$, according to whether $n < M/2$ or $n > M/2$. For rational numbers, this "reconstruction" is not so obvious. Wang [1981] proposed, and Wang *et al.* [1982] justified, using the extended Euclidean algorithm (many other authors made this discovery at about the same time). Each integer calculated in Euclid's algorithm is a linear combination of two data. We apply the algorithm to n and M, to give a decreasing series of integers $a_i = b_i n + c_i M$. In other words, $n \equiv a_i/b_i$ (mod M). It is very remarkable that if there is a fraction a/b equivalent to n with a and b less than $\sqrt{M/2}$, this algorithm will find them, and they will be given by the first element of the series with $a_i < \sqrt{M/2}$. We can adapt Euclid's algorithm to do this calculation.

$$q := M;$$
$$r := n;$$
$$Q := 0;$$
$$R := 1;$$
while $r \neq 0$
        do $t := \text{remainder}(q, r);$
            $T := Q - \lfloor q/r \rfloor R;$
            $q := r;$
            $r := t;$
            $Q := R;$
            $R := T;$
            if $r < \sqrt{M/2}$ **then**
                if $|R| < \sqrt{M/2}$ **then**
                                return r/R
                error "no reconstruction";
        error "common factor";

We have made a simplification with respect to the algorithm of the appendix: it is not necessary to store the dependencies of r with respect to M, and so we do not calculate them in this algorithm.

4.1.3.4 *Other applications*

Of course, we can use modular methods for many calculations linked to the calculation of an inverse, such as the solution of a system of linear equations [Chou and Collins, 1982]: here there is bad reduction when p divides one of the denominators of the solution. Several matrix calculations are described by Gregory and Krishnamurthy [1984] and Krishnamurthy [1985]. As we have already said, Wang [1981] introduced the calculation of fractions to break down a fraction P/QR into partial fractions: here bad reduction corresponds to a value of p such that Q and R are not relatively prime modulo p.

   Here we must describe the great drawback of modular calculation: it does not take into account the sparse nature of a problem. If we have to calculate a polynomial of degree 99, with 100 terms, we need 100 values of the polynomial before we can do the Lagrange interpolation, even if the polynomial has only two non-zero coefficients. We have already said that every realistic polynomial in several variables has to be sparse, and therefore modular methods do not apply very well to these polynomials. Zippel [1979] tried to give a probabilistic character to this method, so as to apply it to the problem of g.c.d. in several variables, and this is the algorithm used by default in MACSYMA.

4.2 *p*-ADIC METHODS

In this section, we shall consider the other large family of advanced methods in Computer Algebra, and we begin with the first application of it — factorisation.

4.2.1 Factorisation of polynomials in one variable

We suppose that there is a polynomial $f(x)$, with integer coefficients, and we ask: can we determine whether there exist polynomials g and h, of degree strictly less than that of f, such that $f = gh$.

Gauss's Lemma *(Equivalent version). Let $f \in \mathbf{Z}[x]$, and $f = gh$ with g and h in $\mathbf{Q}[x]$. Then there is a rational number q such that qg and $q^{-1}h$ belong to $\mathbf{Z}[x]$.*

This lemma, well-known in algebra, indicates that the restriction to factors with integer coefficients does not reduce its generality

Since we can calculate square-free decompositions fairly efficiently, we can suppose that f is square-free. We write $f = a_n x^n + \cdots + a_0$. Every linear factor of f must take the form $bx + c$, where b divides a_n and c divides a_0, and we can therefore look for linear factors. In theory, it is possible to find factors of higher degree by solving systems of equations which determine the coefficients, but this method can only be used if the degrees are very small.

With the success of the modular methods described in the preceding section in mind, we can try to apply them here. It is indeed true that $f = gh$ implies $f_p = g_p h_p$, and, also, that the factorisation modulo p is, at least, a finite problem, for there are only a finite number of possible coefficients. In fact, there is quite an efficient algorithm, which we shall explain in the next sub-section.

4.2.1.1 *Berlekamp's algorithm*

In this section, we suppose that the problem is to factorise a polynomial $f(x)$ modulo p, and f is assumed to be square-free. We must state several facts about calculation modulo p, some of which are well-known.

Proposition 1. *If a and b are two integers modulo p, then $(a + b)^p \equiv a^p + b^p$.*

Proof. By the binomial formula, we can expand $(a + b)^p$ as follows:

$$(a + b)^p = a^p + \binom{p}{1} a^{p-1} b + \cdots + \binom{p}{p-1} ab^{p-1} + b^p$$

$$\equiv a^p + b^p$$

for all the binomial coefficients are divisible by p.

Proposition 2 (Fermat's little theorem). $a^p \equiv a$ *modulo* p.

Proof. By induction on a, for the proposition is true when $a = 0$ or $a = 1$. In general, $a = (a - 1) + 1$, and we can apply the previous proposition to this expression:

$$a^p = ((a - 1) + 1)^p = (a - 1)^p + 1^p = (a - 1) + 1 = a.$$

Corollary. *Every integer modulo p is a root of $x^p - x$, and therefore*

$$x^p - x = (x - 0)(x - 1) \ldots (x - (p - 1)).$$

These two propositions extend to polynomials, in the following manner.

Proposition 3. *If a and b are two polynomials modulo p, then $(a + b)^p \equiv a^p + b^p$.*

The proof is the same as for proposition 1.

Proposition 4. *Let $a(x)$ be a polynomial, then $a(x)^p \equiv a(x^p)$ modulo p.*

Proof. By induction on the degree of a, for the proposition is true when a is only a number, by proposition 2. In general, $a(x) = \hat{a}(x) + a_n x^n$, where $\hat{a}$ is a polynomial of degree less than that of a, and we can apply the previous proposition to this expression:

$$a(x)^p = (\hat{a}(x) + a_n x^n)^p = \hat{a}(x)^p + (a_n x^n)^p = \hat{a}(x^p) + a_n x^{np} = a(x^p).$$

Let us now suppose that f factorises into r irreducible polynomials:

$$f(x) = f_1(x) f_2(x) \ldots f_r(x)$$

(r is unknown for the present). Since f has no multiple factors, the f_i are relatively prime. Let $s_1, \ldots, s_r$ be integers modulo p. By the Chinese remainder theorem (see the appendix "Algebraic background") there is a polynomial v such that

$$v \equiv s_i \quad (\bmod\ p, f_i(x)), \tag{1}$$

where this calculation is modulo the polynomial f_i and modulo the prime number p. Moreover, the degree of v is less than that of the product of the f_i, that is f. Such a polynomial v is useful, for if $s_i \neq s_j$, then $\gcd(f, v - s_i)$ is divisible by f_i, but not by f_j, and therefore leads to a decomposition of f. We have the following relation:

$$v(x)^p \equiv s_j^p \equiv s_j \equiv v(x) \quad (\bmod\ f_j(x), p),$$

and, by the Chinese remainder theorem,

$$v(x)^p \equiv v(x) \quad (\mathrm{mod}\ f(x), p). \tag{2}$$

But, on replacing x by $v(x)$ in the corollary above,

$$v(x)^p - v(x) \equiv (v(x) - 0)(v(x) - 1) \ldots (v(x) - (p-1)) \quad (\mathrm{mod}\ p). \tag{3}$$

Thus, if $v(x)$ satisfies (2), $f(x)$ divides the left hand side of (3), and each of its irreducible factors, the f_i, divides one of the polynomials on the right hand side of (3). But this implies that v is equivalent to an integer modulo f_i, that is that v satisfies (1). We have proved (by Knuth's method [1981], p. 422) the following result.

Berlekamp's Theorem [1967]. *The solutions v of (1) are precisely the solutions of (2).*

We have already said that the solutions of (1) provide information about the factorisation of f, but we still have the problem of finding them. Berlekamp's basic idea is to note that (2) is a *linear* equation for the coefficients of v. This remark may seem strange, but it is a consequence of proposition 4. In fact, if n is the degree of f, let us consider the matrix

$$Q = \begin{pmatrix} q_{0,0} & q_{0,1} & \cdots & q_{0,n-1} \\ q_{1,0} & q_{1,1} & \cdots & q_{1,n-1} \\ \vdots & \vdots & & \vdots \\ q_{n-1,0} & q_{n-1,1} & \cdots & q_{n-1,n-1} \end{pmatrix},$$

where

$$x^{pk} \equiv q_{k,n-1}x^{n-1} + \cdots + q_{k,1}x + q_{k,0} \quad (\mathrm{mod}\ f(x), p).$$

If we consider a polynomial as a vector of its coefficients, multiplication by Q corresponds to the calculation of the p-th power of the polynomial. The solutions of (2) are thus the eigenvectors of the matrix Q (mod p) for the eigenvalue 1. Berlekamp's algorithm can be expressed as follows:

[1] Verify that f has no multiple factors. If not, we have to do a square-free decomposition of f, and apply this algorithm to each of its elements.

[2] Calculate the matrix Q.

[3] Find a basis of its eigenvectors for the eigenvalue 1. One eigenvector is always the vector $[1, 0, 0, \ldots, 0]$, corresponding to the fact that the integers are always solutions of (2). The size of this basis is the number of irreducible factors of f.

[4] Calculate $\gcd(f, v - s)$ for every integer s modulo p, where v is the polynomial corresponding to a non-trivial eigenvector. This ought to

give a decomposition of f. If we find fewer factors than we need, we can use other eigenvectors.

Note that stage [4] is the most expensive, and that, if we stop at stage [3], we have already determined the number of irreducible factors. The running time of this algorithm is $O(n^3 + prn^2)$, where r is the number of factors (average value $\ln n$). This is very fast if p is small, but may be expensive if p is not small. Knuth [1981] describes several variants of this algorithm, some of them where the dependence on p is rather of the form $\log^3 p$.

4.2.1.2 *The modular — integer relationship*

The previous section provides an efficient algorithm for the factorisation modulo p of a polynomial. How can it be used to factorise a polynomial with integer coefficients? Of course, if the polynomial does not factorise modulo p (which does not divide the leading coefficient), the polynomial is then irreducible, for $f = gh$ implies $f_p = g_p h_p$. But the converse is false — for example $x^2 + 1$ is irreducible, but factorises modulo all primes of the form $4k+1$, for -1 is a square modulo a number of that kind. Nevertheless, this polynomial is irreducible modulo numbers of the form $4k+3$, and thus we can prove that it is irreducible over the integers by proving that it is irreducible modulo 3. In fact, there is an infinite number of "good" p, and an infinite number of "bad" p.

Unfortunately, there are irreducible polynomials which factorise modulo all the primes. The simplest example is $x^4 + 1$, which always factorises as the product of two polynomials of degree two, which may still be reducible. This proof requires the use of several properties of squares modulo p — the reader may skip the detailed analysis if he wishes.

$p = 2$ Then $x^4 + 1 = (x+1)^4$.

$p = 4k + 1$ In this case, -1 is always a square, say $-1 = q^2$. This gives us the factorisation $x^4 + 1 = (x^2 - q)(x^2 + q)$.

$p = 8k \pm 1$ In this case, 2 is always a square, say $2 = q^2$. This gives us the factorisation $x^4+1 = (x^2-(2/q)x+1)(x^2+(2/q)x+1)$, which is a version of the factorisation $x^4+4 = (x^2-2x+2)(x^2+2x+2)$ which we met in the section on the representation of simple radicals. In the case $p = 8k + 1$, we have this factorisation and the factorisation given in the previous case. As these two factorisations are not equal, we can calculate the g.c.d.s of the factors, in order to find a factorisation as the product of four linear factors.

$p = 8k + 3$ In this case, -2 is always a square , say $-2 = q^2$. This is a result of the fact that -1 and 2 are not squares, and so their product must be a square. This property of -2 gives us the factorisation $x^4+1 = (x^2 - (2/q)x - 1)(x^2 + (2/q)x - 1)$, which is likewise a version of the factorisation $x^4 + 4 = (x^2 - 2x + 2)(x^2 + 2x + 2)$.

This polynomial is not an isolated oddity: Kaltofen *et al.* [1981, 1983] have proved that there is a whole family of polynomials with this property of being irreducible, but of factorising modulo every prime. Several people have said that these polynomials are, nevertheless, "quite rare", but Abbott *et al.* [1985] have established that they can occur in the manipulation of algebraic numbers.

We have, therefore, to find a method of factorisation which works even if the modular factorisation does not correspond to the factorisation over the integers. At the beginning of this chapter, we quoted the Landau-Mignotte inequality, which lets us bound the the coefficients of a factor of a polynomial, in terms of the degree and size of coefficients of the initial polynomial. This inequality is useful in the present case, for it lets us find an N such that all the coefficients of the factors of a polynomial must lie between $-N/2$ and $N/2$. This gives the following algorithm for factorising the polynomial f, due essentially to Zassenhaus [1969]:

[1] Choose an N sufficiently large for all the coefficients of the factors of f to be smaller than $N/2$. Take care that N does not have a factor in common with the leading coefficient of f.

[2] Factorise the polynomial modulo N. If we suppose f monic, we always choose monic factors. We write $f = f_1 f_2 \ldots f_k \pmod{N}$.

[3] Taking the f_i modulo N as polynomials with integer coefficients, test whether they divide f. Every f_i which divides f is an irreducible factor of f.

[4] If there are any f_i which do not divide f, we have to form polynomials of the type $f_i f_j \pmod{N}$, and test whether these polynomials (with integer coefficients between $-N/2$ and $N/2$) divide f. Every combination which divides f is an irreducible factor of f. Those f_i which occur in a factor of f can be discarded.

[5] If there are some f_i left, we have to form polynomials of the type $f_i f_j f_k \pmod{N}$, and test whether these polynomials (with integer coefficients between $-N/2$ and $N/2$) divide f. Every combination which divides f is an irreducible factor of f. We continue until every combination (or its complement) has been tested.

[6] If there are any f_i left, their product is an irreducible factor of f.

For example, every factor of $x^4 + 1$ must have coefficients less than (or equal to) 16, and therefore it is possible to use $N = 37$, which yields the factorisation $x^4 + 1 \equiv (x^2 - 6)(x^2 + 6) \pmod{37}$, and these factors are irreducible. Considered as polynomials with integer coefficients, these factors do not divide $x^4 + 1$. So we take the combination of them, that is $x^4 + 1$ itself. This divides $x^4 + 1$, and is therefore an irreducible factor, which is what we set out to prove.

This algorithm may be exponential (that is it may require a running time which is an exponential function) in terms of the number of factors found modulo N, because of the number of combinations to be tested in stage [5]. In fact, if the polynomial is irreducible, but has n factors modulo N, there are about 2^{n-1} polynomials which have to be tested to see if they are divisors of f. This problem is called the *combinatorial explosion* of the running cost. It was solved recently by Lenstra *et al.* [1982], but their method seems to be of theoretical rather than of practical interest.

Nevertheless, in practice, we are confronted rather with the following problem: calculating a factorisation modulo N seems to be the most expensive stage of the algorithm. Normally, N is too big for us to be able to take a prime $p > N$, and to use Berlekamp's algorithm directly, for the dependence on p is quite significant, even for the improved versions. Moreover, we want to be able to use small p, so as to find out quickly that a polynomial is irreducible.

Modular methods can be used, but here we come across a problem we have not met before. Suppose, for example, that f factorises as $g_1 h_1$ $\pmod{p_1}$ and as $g_2 h_2$ $\pmod{p_2}$, with all the factors of the same degree. We can apply the Chinese remainder theorem to the coefficients of g_1 and g_2, in order to find a polynomial with coefficients modulo $p_1 p_2$, which is a factor of f modulo $p_1 p_2$ (and similarly for h_1 and h_2). But it is equally possible to apply the theorem to g_1 and h_2, or to g_2 and h_1. In effect, the ring $(\mathbf{Z}/p_1 p_2 \mathbf{Z})[x]$ does not have unique factorisation, and f has two different factorisations modulo $p_1 p_2$. If we find the same behaviour modulo p_3, we find four factorisations modulo $p_1 p_2 p_3$, and, in general, the number of different factorisations may be an exponential function of the number of p chosen. Thus, the modular approach is not very useful for this calculation, and we have to find a different method for factorising modulo N. This time we shall use $N = p^k$, where p is a prime, and such methods are called p-adic.

4.2.2 Hensel's Lemma — linear version.

In this section, we shall see how to calculate a factorisation of f modulo p^k, starting from a factorisation modulo p calculated by Berlekamp's algorithm, which we have already described. For simplicity, we consider first the case of a monic polynomial f, which factorises modulo p as $f = gh$, where g and h are relatively prime (which implies that f modulo p is square-free, that is that p does not divide the resultant of f and f'). We use subscripts to indicate the power of p modulo which an object has been calculated. Thus our factorisation can be written $f_1 = g_1 h_1 \pmod{p^1}$ and our aim is to calculate a corresponding factorisation $f_k = g_k h_k \pmod{p^k}$ such that p^k is sufficiently large.

Obviously, $g_2 \equiv g_1 \pmod{p}$, and therefore we can write $g_2 = g_1 + p\hat{g}_2$ where $\hat{g}_2$ is a measure of the difference between g_1 and g_2. The same holds for f and h, so that $f_2 = g_2 h_2 \pmod{p^2}$ becomes

$$f_1 + p\hat{f}_2 = (g_1 + p\hat{g}_2)(h_1 + p\hat{h}_2) \pmod{p^2}.$$

Since $f_1 = g_1 h_1 \pmod{p^1}$, this equation can be rewritten in the form

$$\frac{f_1 - g_1 h_1}{p} + \hat{f}_2 = \hat{g}_2 h_1 + \hat{h}_2 g_1 \pmod{p}.$$

The left hand side of this equation is known, whereas the right hand side depends linearly on the unknowns $\hat{g}_2$ and $\hat{h}_2$. Applying the extended Euclidean algorithm (see the appendix "Algebraic background") to g_1 and h_1, which are relatively prime, we can find polynomials $\hat{g}_2$ and $\hat{h}_2$ of degree less than g_1 and h_1 respectively, which satisfy this equation modulo p. The restrictions on the degrees of $\hat{g}_2$ and $\hat{h}_2$ are valid in the present case, for the leading coefficients of g_k and h_k have to be 1. Thus we can determine g_2 and h_2.

Similarly, $g_3 \equiv g_2 \pmod{p^2}$, and we can therefore write $g_3 = g_2 + p^2 \hat{g}_3$ where $\hat{g}_3$ is a measure of the difference between g_2 and g_3. The same is true for f and h, so that $f_3 = g_3 h_3 \pmod{p^3}$ becomes

$$f_2 + p^2 \hat{f}_3 = (g_2 + p^2 \hat{g}_3)(h_2 + p^2 \hat{h}_3) \pmod{p^3}.$$

Since $f_2 = g_2 h_2 \pmod{p^2}$, this equation can be rewritten in the form

$$\frac{f_2 - g_2 h_2}{p^2} + \hat{f}_3 = \hat{g}_3 h_2 + \hat{h}_3 g_2 \pmod{p}.$$

Moreover, $g_2 \equiv g_1 \pmod{p}$, so this equation simplifies to

$$\frac{f_2 - g_2 h_2}{p^2} + \hat{f}_3 = \hat{g}_3 h_1 + \hat{h}_3 g_1 \pmod{p}.$$

The left hand side of this equation is known, whilst the right hand side depends linearly on the unknowns $\hat{g}_3$ and $\hat{h}_3$. Applying the extended Euclidean algorithm to g_1 and h_1, which are relatively prime, we can find the polynomials $\hat{g}_3$ and $\hat{h}_3$ of degrees less than those of g_1 and h_1 respectively, which satisfy this equation modulo p. Thus we determine g_3 and h_3 starting from g_2 and h_2, and we can continue these deductions in the same way for every power p^k of p until p^k is sufficiently large.

We must note that Euclid's algorithm is always applied to the same polynomials, and therefore it suffices to perform it once. In fact, we can state the algorithm in the following form:

> **Algorithm** *Hensel's — linear version*;
> **Input** f, g_1, h_1, p, k;
> **Output** g_k, h_k;
> $g := g_1$;
> $h := h_1$;
> $grecip, hrecip := ExtendedEuclid(g_1, h_1, p)$;
> **for** $i := 2 \ldots k$ **do**
> $$Discrepancy := \frac{f - gh \pmod{p^i}}{p^{i-1}};$$
> $gcorr := Discrepancy * hrecip \pmod{p, g_1}$;
> $hcorr := Discrepancy * grecip \pmod{p, h_1}$;
> $g := g + p^{i-1}gcorr$;
> $h := h + p^{i-1}hcorr$;
> **return** g, h;

Here the algorithm *ExtendedEuclid* has to return *grecip* and *hrecip* such that

$$grecip * g_1 + hrecip * h_1 = 1 \pmod{p}.$$

4.2.2.1 *Hensel's Lemma — quadratic version*

There is another version of this algorithm, which doubles the exponent of p at each stage, rather than increasing it by one. We shall give the mathematical explanation of the move from p^2 to p^4, and then the general algorithm. In fact, $g_4 \equiv g_2 \pmod{p^2}$, and therefore we can write $g_4 = g_2 + p^2 \breve{g}_4$ where $\breve{g}_4$ is a measure of the difference between g_2 and g_4. The same is true of f and h, so that $f_4 = g_4 h_4 \pmod{p^4}$ becomes

$$f_2 + p^2 \breve{f}_4 = (g_2 + p^2 \breve{g}_4)(h_2 + p^2 \breve{h}_4) \pmod{p^4}.$$

Since $f_2 = g_2 h_2 \pmod{p^2}$, this equation can be rewritten in the form

$$\frac{f_2 - g_2 h_2}{p^2} + \breve{f}_4 = \breve{g}_4 h_2 + \breve{h}_4 g_2 \pmod{p^2}.$$

The left hand side of this equation is known, whilst the right hand side depends linearly on the unknowns $\breve{g}_4$ and $\breve{h}_4$. By applying the extended

Euclidean algorithm to g_2 and h_2, which are relatively prime modulo p^2 because they are so modulo p, we can find polynomials $\breve{g}_4$ and $\breve{h}_4$ of degrees less than those of g_2 and h_2 respectively, satisfying this equation modulo p^2. The restrictions on the degrees of $\breve{g}_4$ and $\breve{h}_4$ hold in the present case, because the leading coefficients of g and h have to be 1. So we have determined g_4 and h_4 directly, starting from g_2 and h_2, we can continue in the same way for every power of p with exponent a power of two until p^{2^l} is sufficiently large.

> **Algorithm** *Hensel — quadratic version 1*;
> **Input** f, g_1, h_1, p, k;
> **Output** $g_k, h_k, modulus$;
> $g := g_1$;
> $h := h_1$;
> $modulus := p$;
> **for** $i := 1 \ldots \lceil \log_2 k \rceil$ **do**
> $\qquad Discrepancy := \dfrac{f - gh \quad (\text{mod } modulus^2)}{modulus}$;
> $\qquad grecip, hrecip := ExtendedEuclid(g, h, modulus)$;
> $\qquad gcorr := Discrepancy * hrecip \quad (\text{mod } modulus, g)$;
> $\qquad hcorr := Discrepancy * grecip \quad (\text{mod } modulus, h)$;
> $\qquad g := g + modulus * gcorr$;
> $\qquad h := h + modulus * hcorr$;
> $\qquad modulus := modulus^2$;
> **return** $g, h, modulus$;

This algorithm gives solutions modulo a power of p which may be greater than p^k, but which can always be reduced if necessary. Here the algorithm *ExtendedEuclid* must give *grecip* and *hrecip* such that

$$grecip * g + hrecip * h = 1 \quad (\text{mod } modulus).$$

We can look at Hensel's methods as variations of another well-known method with quadratic convergence — that of Newton for finding the simple roots of real equations. This method starts from an approximation x_1 to the root of $F(x) = 0$, and calculates successively better approximations by the sequence

$$x_{2n} = x_n - \frac{F(x_n)}{F'(x_n)}.$$

In general, x_{2n} has twice as many correct digits as x_n has. If we take y as a variable, and $F = f - gh$, then we find $g_{2n} = g_n + \frac{f - g_n h}{h}$, which is the equation we use. There is, in fact, a theory of *p-adic numbers*, where two numbers are "close" if they differ from one another by a multiple of a power of p (the greater the power, the closer the integers). With this idea of "distance", Hensel's quadratic algorithm is precisely Newton's algorithm, and the need for a simple root implies that g and h have to be relatively prime. Lipson [1976] studies this connection.

This quadratic convergence is very useful in the case of real numbers, but it is less helpful in the present case. For floating-point numbers, in general, the running time is independent of the data, whereas, for us, precision is equivalent to the length of the integers, and therefore a doubling of precision corresponds to a multiplication of the running time by four. Comparison of linear and quadratic methods depends on details of implementation, and there is no clear cut result in this area. Several authors use a hybrid method, that is they use the quadratic method as long as p^i is contained in one word in the computer, and then they go over to the linear method.

There are several versions of Hensel's method: we shall describe only some of them. First, we note that it is possible to work with several factors at once, rather than with two factors. It suffices to calculate the inverses of all the factors (except g!) modulo g for each factor g. In the linear algorithm, this is not very costly, for the inverses are only calculated once, but it is quite costly for the quadratic algorithm.

4.2.2.2 *Hensel's Lemma — refinement of the inverses*

This observation has led several people to note that the inverses modulo p^k are related to the inverses modulo $p^{k/2}$: in fact they are equivalent modulo $p^{k/2}$. We can see the inverse G of g as a solution of the equation $Gg \equiv 1 \pmod{h, p^k}$, and this equation can be solved by Hensel's method. As before, we shall study the move from p^2 to p^4, by supposing that $g_2 G_2 \equiv 1 \pmod{h_2, p^2}$, and that the aim is to find a G_4 such that $g_4 G_4 \equiv 1 \pmod{h_4, p^4}$. We write $g_4 = g_2 + p^2 \breve{g}_4$ as before, and $G_4 = G_2 + p^2 \breve{G}_4$. Thus the equation we want is translated into

$$(g_2 + p^2 \breve{g}_4)(G_2 + p^2 \breve{G}_4) = 1 \pmod{h_4, p^4}.$$

This is rewritten in the form

$$p^2 g_2 \breve{G}_4 = 1 - g_2 G_2 - p^2 \breve{g}_4 G_2 \pmod{h_4, p^4}.$$

By the recursive definition of G_2, p^2 divides the right hand side of this equation, and therefore we find

$$g_2 \breve{G}_4 = \frac{1 - g_2 G_2}{p^2} - \breve{g}_4 G_2 \pmod{h_4, p^2}.$$

But $h_2 = h_4 \pmod{p^2}$, and therefore this equation is effectively modulo h_2 and p^2. But, by induction, G_2 is the inverse of g_2, and this equation can be solved for $\check{G}_4$:

$$\check{G}_4 = G_2 \left(\frac{1 - g_2 G_2}{p^2} - \check{g}_4 G_2 \right) \pmod{h_2, p^2}.$$

This equation gives us a Hensel algorithm which only uses the extended Euclidean algorithm at the beginning. We state it, as usual, for the case of two factors.

Algorithm *Hensel — quadratic version 2*;
Input f, g_1, h_1, p, k;
Output $g_k, h_k, modulus$;
$g := g_1$;
$h := h_1$;
$grecip, hrecip := ExtendedEuclid(g_1, h_1, p)$;
$modulus := p$;
for $i := 1 \ldots \lceil \log_2 k \rceil$ **do**

$\qquad Error := \dfrac{f - gh \pmod{modulus^2}}{modulus}$;

$\qquad gcorr := Error * hrecip \pmod{modulus, g}$;
$\qquad hcorr := Error * grecip \pmod{modulus, h}$;
$\qquad g := g + modulus * gcorr$;
$\qquad h := h + modulus * hcorr$;
$\qquad$ **if** this is not the last iteration

$\qquad\qquad$ **then** $Error := \dfrac{1 - g * grecip \pmod{h, modulus^2}}{modulus} - gcorr * grecip$;

$\qquad\qquad grecipcorr := Error * grecip \pmod{modulus, h}$;

$\qquad\qquad Error := \dfrac{1 - h * hrecip \pmod{g, modulus^2}}{modulus} - hcorr * hrecip$;

$\qquad\qquad hrecipcorr := Error * hrecip \pmod{modulus, g}$;
$\qquad\qquad grecip := grecip + modulus * grecipcorr$;
$\qquad\qquad hrecip := hrecip + modulus * hrecipcorr$;
$\qquad modulus := modulus^2$;
return $g, h, modulus$;

4.2.2.3 *The factorisation algorithm*

Using the Hensel algorithm from the previous section, we can give an algorithm for factorising polynomials with integer coefficients. First, we suppose that our polynomial f is monic and square-free (if not, we first find the square-free decomposition by the method described in the appendix "Algebraic background").

Algorithm for factorising in one variable

do $p := prime()$;
   $f_p := f \pmod{p}$;
    while $degree(f) \neq degree(f_p)$ **or** $\gcd(f_p, f_p') \neq 1$;
  $\{g_1, \ldots, g_n\} := Berlekamp(f_p, p)$;
  if $n = 1$ **then return** $\{f\}$;
  $k := \log_p(2 * Landau\_Mignotte\_bound(f))$;
  $Factors := Hensel(f, \{g_1, \ldots, g_n\}, p, k)$;
  $Answer := \{\}$;
  for each element g of $Factors$ **do**
   if g divides f **then**
     $Answer := Answer \cup \{g\}$;
     $Factors := Factors \setminus \{g\}$;
     $n := n - 1$;
  $Combine := 2$;
  while $2 * Combine \leq n$ **do**
   for each $Combine$-subset E of $Factors$ **do**
   $g := \prod_{h \in E} h \pmod{p^k}$;
   if g divides f **then**
     $Answer := Answer \cup \{g\}$;
     $Factors := Factors \setminus E$;
     $n := n - Combine$;
     if $2 * Combine > n$ **then**
      exit from both loops;
   $Combine := Combine + 1$;
  if $Factors \neq \emptyset$ **then**
$$g := \prod_{h \in Factors} h \pmod{p^k}$$
   $Answer := Answer \cup \{g\}$;
  return $Answer$;

In this algorithm we suppose that the sub-algorithm *prime* returns a different prime number at each call, for example by running through the list $2, 3, 5, 7, 11, \ldots$. *Berlekamp* applies the Berlekamp algorithm described in a previous sub-section (or one of the variants of it), and returns a list of the factors of f_p. *Landau_Mignotte_bound* returns an integer such that each factor of f has all its coefficients less than this integer. We have to multiply by two to allow for the choice of signs. We suppose that there is a Hensel algorithm (called *Hensel*) which takes n factors, although we have only given a version which takes two. We are not interested here in the details (linear/quadratic etc.).

There are several possible optimisations. For example, if $2 * Combine = n$, there is no need to look at all the *Combine*-subsets of *Factors*, but only at half of them, for the others are their complements.

It would be possible to take several values of p, in order to find one which gives the minimum number of modular factors. In an even more complicated way, we can try to get the maximum information from factorisations modulo the different p. For example, if f factorises into two polynomials of degree two modulo a prime, and a polynomial of degree three and one of degree one modulo another prime, we can deduce that f is irreducible. But programming these tests is not self-evident. Musser [1978] claims that the optimal number of p to choose is five, but the exact value of this number obviously depends on the implementation of the algorithms.

Note that the expensive stage is, in practice, the Hensel lemma, although, in theory, trying all the combinations may take an exponential time. If we have to test several combinations, the division test may be very costly. Abbott *et al.* [1985] have pointed out that it is possible to improve this stage by testing that the trailing coefficients divide before testing the polynomials.

The algorithm we have described is quite well structured, with one main program and two sub-programs, that is *Berlekamp* and *Hensel*. We must confess that production implementations are much more complicated than this ideal structure would suggest.

(a) Berlekamp's algorithm gives the number of factors (that is the size of the basis of eigenvectors) after the third stage, whilst many more calculations are needed to find the factors. It is therefore possible to start several (five according to Musser [1978]) instances of Berlekamp's algorithm with different p, and to stop them after the third stage, get all possible information from the numbers of factors, and then to restart again the promising instances. This is easy to say, but more difficult to write in most existing programming languages.

(b) It is possible to test, during the Hensel algorithm, whether one of the factors (mod p^j, or p^{2^j} for a quadratic version) is a true factor of the polynomial [Wang, 1983]. We can use the method described for the modular g.c.d., and test one factor to see if it changes after one Hensel stage, that is we can add, after lines of the form

$$gcorr := Error * hrecip \quad (\bmod \ modulus, g);$$

lines of the style

$$\textbf{if } gcorr = 0$$
$$\textbf{then if } g \text{ divides } f \textbf{ then} \dots;$$

If this happens, we can remove it from the list of the p-adic factors and reduce f. This may change the Landau-Mignotte bound, and therefore the necessary power of p. This optimisation means that most of the variables of the factorisation algorithm can be modified by Hensel's algorithm: a violation of the principles of modular programming.

These remarks, and others we have made, imply that real implementations are quite complicated programs.

4.2.2.4 *The leading coefficient*

In the previous sections we have supposed that the polynomial we want to factorise is monic. Every problem of factorising polynomials can be reduced to this case, by performing the following two steps:

[1] Multiply the polynomial f by a_n^{n-1}, where n is the degree of f and a_n is its leading coefficient.

[2] Make the substitution $y = x/a_n$.

Since the leading coefficient is now a_n^n, and all the other coefficients are divisible by a_n^{n-1}, this substitution makes the polynomial monic, and does not take it out of $\mathbf{Z}[x]$. In fact, the polynomial $\sum_{i=0}^{n} a_i x^i$ is replaced by the polynomial

$$\sum_{i=0}^{n} a_i a_n^{n-1-i} x^i = x^n + a_{n-1} x^{n-1} + a_n a_{n-2} x^{n-2} + \cdots + a_n^{n-1} a_0.$$

This substitution has the great disadvantage of making the coefficients, and therefore the Landau-Mignotte bound, grow, and thus the size of all the integers entering into the algorithm increases. Is there a different method?

We recall that we could only consider monic f because of the condition on the degrees of the "corrections" $\hat{g}_i$, which were always smaller in degree than that of g_i. Indeed, a factorisation of a polynomial of $R[x]$ is only determined to within units (invertible elements of R). For integers modulo p, each non-zero element is invertible, and therefore it is possible to multiply the factorisation by any number. The basic idea is to make the polynomial monic modulo p at the outset, and to do the division by using the following remark: if g divides f, then cg divides $a_n f$, where the constant c is chosen such that the leading coefficient of cg will be a_n. Thus the algorithm has to be modified as follows:

(a) the *Berlekamp* call has to be on the monic version of f_p;

(b) the Landau-Mignotte bound has to be multiplied by a_n, because it is now the coefficients of $a_n g$ which have to be bounded (rather than those of g);

(c) in the Hensel algorithm we have to replace f by f/a_n (calculated modulo a suitable power of p);

(d) the division tests of the kind

$$\textbf{if } g \text{ divides } f$$

have to be replaced by tests of the following kind:

$$\textbf{if } a_n g \pmod{p^k} \text{ divides } a_n f$$

and it is the primitive part of $a_n g \pmod{p^k}$ which has to be added to the result.

We can also use the determination of rational numbers starting from their modular values, as described in the sub-section "Inverse of a matrix" (4.1.3.3). Instead of factorising a polynomial with integer coefficients, we can factorise a monic polynomial with rational coefficients. Of course, p must not divide the leading coefficient of the initial polynomial, which becomes the denominator of the rational coefficients. Wang [1983] gives an account of this.

4.2.3 Factorisation in several variables

The factorisation of polynomials in several variables is done in the same way as factorisation in one variable. As in the case of a g.c.d., the idea of reduction modulo p is replaced by the idea of an evaluation $y := r$, and the variables are replaced until we find a polynomial in one variable. We shall only give a brief account of this algorithm, because the details are complicated, even though the principles are similar to those we have already seen.

We shall need, of course, an equivalent of Hensel's lemma. This is quite easy to construct: we replace p by $y - r$ everywhere in the equations and the algorithms. There are linear and quadratic versions of these algorithms, and it is not obvious which is the best, but it seems that most systems use a linear version.

We do not need a Landau-Mignotte style inequality, for the degree of a factor is always less than the degree of the initial polynomial. Moreover, we can say that, if the degree of a polynomial is n, at least one of its factors is of degree less than or equal to $n/2$.

There is still the same idea of bad reduction. For example, the polynomial $x^2 - y$ is irreducible (because its degree in y is one), but it factorises at each square value of y, such as $y = 1$ or $y = 4$. Now the situation is very much better than in the case of polynomials in one variable.

Hilbert's Irreducibilty Theorem. *Let $f(y_1, \ldots, y_s, x_1, \ldots, x_r)$ be an irreducible polynomial of $\mathbf{Z}[y_1, \ldots, y_s, x_1, \ldots, x_r]$. Let $M(N)$ be the number of evaluations $y_1 = b_1, \ldots, y_s = b_s$ with $|b_i| < N$ such that the polynomial $f(b_1, \ldots, b_s, x_1, \ldots, x_r)$ of $\mathbf{Z}[x_1, \ldots, x_r]$ is reducible, that is that the evaluation is bad. Then there exist constants α and C with $0 < \alpha < 1$ and $M(N) \leq C(2N+1)^{s-\alpha}$.*

As the number of possible evaluations, written $P(N)$, is $(2N+1)^s$, this theorem states that $\lim_{N \to \infty} M(N)/P(N) = 0$: in other words, that the probability that an evaluation will be "bad" tends towards zero when the the integers become sufficiently large. In practice, bad reductions seem to be quite rare.

4.2.3.1 *The algorithm*

We still suppose that the polynomial to be factorised is square-free. If the variables of this polynomial f are $x_1, \ldots, x_r$, we shall describe an algorithm which has as parameters f, r and s, the latter being the the index of the variable which is to be replaced, supposing that $x_{s+1}, \ldots, x_{r-1}$ have already been replaced by integer values. Suppose, to start with, that the polynomial f to be factorised is monic with respect to x_r.

We have given a version which only raises the factors to a power of $(x_s - v)$ such that one of the two factors has to be of degree less than or equal to this power. This decision implies that it is necessary to test all the combinations of factors, rather than half of them, but this seems to be more efficient than to go up to twice the degree, supposing that bad reductions are quite rare.

In this algorithm we suppose that *random* gives us a random value for the evaluation, choosing first the smallest values. We use *factorise-univariate* to factorise the polynomials in one variable, and the *Hensel* sub-algorithm to apply Hensel's lemma which raises a factorisation modulo $x_s - v$ up to $(x_s - v)^k$.

The algorithm as presented here is recursive in the variables: we can imagine a parallel version which evaluates all the variables but one, factorises this polynomial, and then does several Hensel-type parallel stages to recover the variables. The parallel versions are in general more efficient, but more complicated to describe (and to write!). For a description see the article by Wang [1978].

There are three possibilities which may increase the running time needed for this algorithm, which is in general quite fast. Classically they are described (for example, by Wang [1978]) as being:

(1) **bad zeros**, that is the impossibility of using 0 as value v;

(2) **parasitic factors**, that is bad reduction;

Algorithm of factorisation in several variables

proc *factorise*(f, r, s);
    if $s = 0$ **then return** *factorise-univariate*(f);
    do $v := random()$;
        $fv := f_{x_s := v}$;
        while $degree(f) \neq degree(fv)$ **or** $\gcd(fv, fv') \neq 1$;
    $\{g_1, \ldots, g_n\} := factorise(fv, r, s - 1)$;
    if $n = 1$ **then return** $\{f\}$;
    $k := degree(f, x_s)/2$;
    $Factors := Hensel(f, \{g_1, \ldots, g_n\}, x_s - v, k)$;
    $Answer := \{\}$;
    for each element g **of** *Factors* **do**
        if g divides f **then**
                $Answer := Answer \cup \{g\}$;
                $Factors := Factors \setminus \{g\}$;
                $n := n - 1$;
    $Combine := 2$;
    while $Combine < n$ **do**
        for each $Combine$-subset E **of** *Factors* **do**
           $g := \prod_{h \in E} h \quad (\mathrm{mod}\ (x_s - v)^k)$;
        if g divides f **then**
                $Answer := Answer \cup \{g\}$;
                $Factors := Factors \setminus E$;
                $n := n - Combine$;
        $Combine := Combine + 1$;
    if $Factors \neq \emptyset$ **then**
        $$g := \prod_{h\ \in\ Factors} h \quad (\mathrm{mod}\ (x_s - v)^k);$$
        $Answer := Answer \cup \{g\}$;
    return *Answer*;

(3) **leading coefficient**, which we have already seen in the case of one variable.

Bad zeros are very serious in practice, because of a phenomenon we have already come across in the section "Polynomials in several variables", that is that every realistic polynomial in several variables must be sparse. A sparse polynomial remains sparse modulo x_s^n, but becomes dense (in the variable x_s) modulo $(x_s - v)^n$ for non-zero v. In fact, evaluation at a non-zero value corresponds to the substitution $x_s \rightarrow x_s - v$ followed by an evaluation $x_s = 0$, and we have already noted that such a substitution destroys the sparse nature of a polynomial, and can greatly increase the running time. This is not very serious for the recursive method, but is disastrous for the parallel method. There are several devices which try to avoid this problem, and we refer the reader to the papers by Viry [1982] and Wang [1978] for the details.

The effects of parasitic factors can be mitigated by taking several evaluations, as we have already seen in the case of one variable. But to do this becomes recursively very expensive: if we take two evaluations instead of one, the cost of the calculation is multiplied by 2^{r-1}, and it is probable that we shall need many more than two. Bad evaluations give rise to a *combinatorial explosion*, which we have already come across in the case of one variable, but here they cause another problem as well. It is very probable that the false factors are not sparse, and therefore that the Hensel lemma will be very expensive to execute.

4.2.3.2 *The leading coefficient*

We have already seen in the case of one variable that it is possible to remove the leading coefficient by a change of variable. But this was rather expensive in that case, and in the present case it would indeed be very expensive, because it increases the degree and destroys the sparse nature. For example, let us suppose that we have to factorise a polynomial of degree five in four variables. If the leading coefficient is of degree five, the degree after substitution would be 25, and a fairly small problem has turned into a very big problem. As for the number of terms, the polynomial

$$(w^4 + x^4 + y^4 + 1)z^4 + (w + x + y + 1)(z^4 + 4^3 + z^2 + z + 1)$$

contains 24 terms (in its expanded form), but its equivalent which is monic with respect to z contains 247 (and fills several screens).

We have suggested a different method for the case of one variable, which consists in working with the monic equivalent (modulo p^k) of the polynomial to be factorised. This is not very efficient in the present case, for the inverse (modulo $(x_s - v)^k$) of a polynomial can be quite large, even if it can be calculated (for the ring of the other variables is not necessarily a field, and the inverse may well require rational fractions in the other variables $x_1, \ldots, x_{s-1}$). So we have to find another device.

Wang [1978] found one in a parallel setting, that is where the polynomial is reduced to one variable. We put

$$f(x_1, \ldots, x_r) = a_n(x_1, \ldots, x_{r-1})x_r^n + \ldots,$$

where a_n is the leading coefficient troubling us. Wang suggests finding an evaluation $v_1, \ldots, v_{r-1}$ of the variables $x_1, \ldots, x_{r-1}$ which satisfies the following three conditions:

(1) $a_n(v_1, \ldots, v_{r-1})$ does not cancel — this is necessary so that the degree of the reduced polynomial is equal to the degree of the initial polynomial;

(2) $f(v_1, \ldots, v_{r-1}, x_r)$ has no multiple factors — this is necessary for the Hensel lemma;

(3) each factor (after suppressing multiple factors) of the leading coefficient $a_n(x_1, \ldots, x_{r-1})$ (these are all polynomials with $r-1$ variables), when evaluated at $x_1 = v_1, \ldots, x_{r-1} = v_{r-1}$, contains as factor a prime which is not contained in the evaluations of the other factors.

The first two conditions are always necessary — it is the third which is the key to Wang's method. When we have found a substitution of this kind, we can factorise the leading coefficients of all the factors of $f(v_1, \ldots, v_{r-1}, x_r)$. When we look at the primes which identify the factors of a_n, we can determine the leading coefficients of those factors in n variables which correspond to those factors in one variable.

As soon as we know the leading coefficients, we can insist on these leading coefficients for the factors before starting Hensel's algorithm. In this case, all the corrections satisfy the condition on their degrees which is fundamental to this algorithm. Therefore, we have a method of factorising which can work well with a non-trivial leading coefficient, and which does not require any transformation of the polynomial. The cost of this method is the sum of two components:

(1) the factorisation of the leading coefficient (and, by induction, its leading coefficient etc.);

(2) the difficulty of finding substitutions which satisfy all these conditions. The second component seems to be the more significant, but research is in progress in this area.

4.2.4 Other applications of *p*-adic methods

For a *p*-adic method to be valid, we have to have a Hensel algorithm, that is an algorithm for going from the solution modulo p^k to the solution modulo p^{k+1} or p^{2k} (in the case of several variables, we replace p by $x_s - v$). We have mentioned two algorithms of this kind — the refinement of the factors of a polynomial, and the refinement of multiplicative inverses. Wang [1981] has generalised the application to the inverses for the calculation of decompositions into partial fractions.

As an example of another application of refinement of factors, we now explain the *p*-adic method for calculating g.c.d.s. And we can compare the advantages and disadvantages of the modular and *p*-adic methods.

4.2.4.1 *g.c.d. by a p-adic algorithm.*

Let us first consider the case of one variable, and, as at the beginning of this chapter, let us suppose that A and B are two polynomials with integer coefficients of which we want to calculate the g.c.d. We choose a prime p which does not divide the two leading coefficients of A and B, and let F_1 be the g.c.d. of A (mod p) and B (mod p) (we use subscripts to denote the power of p modulo which we are calculating). Thus we can write the following system:

$$A_1 \equiv F_1 G_1 \quad (\text{mod } p); \qquad\qquad (A)$$
$$B_1 \equiv F_1 H_1 \quad (\text{mod } p). \qquad\qquad (B)$$

For the present, we consider the case where $\gcd(A, B)$ is monic (for example, A or B is monic, or, more generally, the g.c.d. of their leading coefficients is one). Always supposing that the reduction was good, we have to refine one of these equations until the power of p is sufficiently great for F_k to be equal to the true solution F. This can indeed be done if F_1 and G_1 are relatively prime, by applying Hensel's lemma to equation (A). Or, if F_1 and H_1 are relatively prime, we can apply Hensel's algorithm to equation (B). What can be done if neither of these possibilities holds?

There are two solutions.

(a) We can do one step of Euclid's algorithm before going over to the p-adic methods. Instead of considering A and B, we consider B and $C = \text{remainder}(A, B)$ (supposing that A is of degree greater than that of B). By Euclid's algorithm, $\gcd(A, B) = \gcd(B, C)$. Likewise $\gcd(A_1, B_1) = \gcd(B_1, C_1)$. None of the factors of $\gcd(F_1, G_1)$ or $\gcd(F_1, H_1)$ can divide $\gcd(F_1, C_1/F_1)$, and therefore, after a finite number of these steps, we have exhausted all the factors of F_1, and one of the conditions of Hensel's algorithm is then necessarily satisfied.

(b) We can take a linear combination of A and B, that is $D = \lambda A + \mu B$, and take $J_1 = \lambda G_1 + \mu H_1$. With probability one, F_1 and J_1 are relatively prime, and we have refined the equation

$$D_1 = F_1 J_1.$$

If we cannot suppose that the true g.c.d. is monic, we can follow the method used for factorising in one variable, by multiplying the equation throughout by a bound for the leading coefficient, that is the g.c.d. of the leading coefficients of A and B. In any event, we arrive at a refinement of the form $A_k = F_k G_k$. If F_k divides A (seen as polynomials with integer coefficients), then F_k is the g.c.d. of A and B. If not, the reduction was bad, and we have to start again. From this we can deduce one disadvantage of the p-adic method — it is more sensitive to bad reductions. This defect can be mitigated by looking for several reductions, for a good reduction will give the lowest degree.

These methods can be generalised to polynomials in several variables: see Yun [1974]. Here we can see one advantage of the p-adic method — it can take advantage of the sparse nature of the data. To be able to use it, we have to overcome the three problems of bad zeros, parasitic factors and leading coefficient. Wang [1980] has suggested ways of approaching these problems, rather similar to those we have outlined for factorisation.

5. Formal integration and differential equations

5.1 FORMAL INTEGRATION

In this section we shall describe the theory and practice of formal integration. Computers are very useful for numerical integration, that is the finding of definite integrals. But Computer Algebra also lets us perform formal integration, that is the discovery of integrals as formulae. These calculations, which cannot be done numerically, are one of the great successes of Computer Algebra.

5.1.1 Introduction

We must distinguish between formal integration and numerical integration. Naturally, numerical integration which consists in giving a numerical value to a definite integral, such as $\int_0^1 x^2 \, dx = 1/3$, was one of the first ways the computer was used. Formal differentiation was undertaken quite early in the history of computers [Kahrimanian, 1953; Nolan, 1953], but it was Slagle [1961] who took the first steps towards integration. Why was there this delay?

The reason is to be found in the big difference between formal integration and formal differentiation. Differentiation, as it is taught in school, is an algorithmic procedure, and a knowledge of the derivatives of functions plus the following four rules

$$(a \pm b)' = a' \pm b'$$
$$(ab)' = a'b + ab'$$
$$\left(\frac{a}{b}\right)' = \frac{a'b - ab'}{b^2}$$
$$f(g(t))' = f'(g(t))g'(t)$$

enables us to differentiate any given function. In fact, the real problem in differentiation is the simplification of the result, because if it is not simplified, the derivative of $2x + 1$ is given as $0x + 2 * 1 + 0$, or, if $2x + 1$ is represented as $2x^1 + 1x^0$,

$$0x^1 + 2\left(0(\log x)x^1 + 1 * 1 * x^{1-1}\right) + 0x^0 + 1\left(0(\log x)x^0 + 0 * 1 * x^{0-1}\right).$$

On the contrary, integration seems to be a random collection of devices and of special cases. It looks as if there is only one general rule, $\int f + g = \int f + \int g$ (and this rule is not always valid — see equation (1) in the section "Integration of rational fractions" and the subsequent discussion). For combinations other than addition (and subtraction) there are no general rules. For example, because we know how to integrate $\exp x$ and x^2 it does not follow that we know how to integrate their composition $\exp x^2$: as we shall see later, this function has no integral simpler than $\int \exp x^2 dx$. So we learn several "methods" such as: integration by parts, integration by substitution, integration by looking up in the printed tables of integrals etc. And in addition we do not know which method or which combination of methods will work for a given integral. So the first moves (made by Slagle) were based on the same heuristics as those used by humans.

This way was quite quickly outdated (after the work of Moses [1967; 1971b]) by truly algorithmic methods. Now there is a whole theory of integration, which we can only summarise quite briefly. Unless otherwise indicated, all integration in this chapter is with respect to x.

Since differentiation is definitely simpler than integration, it is appropriate to rephrase the problem of integration as the "inverse problem" of differentiation, that is, given a function a, instead of looking for its integral b, we ask for a function b such that $b' = a$.

Definition. *Given two classes of functions A and B, the integration problem from A to B is to find an algorithm which, for every member a of A, either gives an element b of B such that $a = b'$, or proves that there is no element b of B such that $a = b'$.*

For example, if $A = \mathbf{Q}(x)$ and $B = \mathbf{Q}(x)$, then the answer for $1/x^2$ must be $-1/x$, whilst for $1/x$ it must be "impossible". On the other hand, if $B = \mathbf{Q}(x, \log x)$, then the answer for $1/x$ must be $\log x$.

Richardson [1968] gives a theorem, which proves that the problem of integration for $A = B = \mathbf{Q}(i, \pi, \exp, \log, ||)$ (where $||$ denotes the absolute value) is insoluble. But those wanting an integration algorithm need not despair: in fact the problem of determining whether a constant is zero or not cannot be decided within this field. So, Richardson takes an indeterminable constant c, and considers the function ce^{x^2}. Since, as is well known (and as we shall prove later), the function e^{x^2} cannot be integrated in B, we see that ce^{x^2} can only be integrated if $c = 0$, a question which cannot be decided. From now on, we shall suppose that our classes of functions are *effective*, that is that every problem of equality can be decided.

5.1.2 Integration of rational functions

In this section, we deal with the case of $A = C(x)$, where C is a field of constants (and, according to what we have just said, effective). Every rational function f can be written in the form $p + q/r$, where p, q and r are polynomials, q and r are relatively prime, and the degree of q is less than that of r. It is well known that

$$\int x + y = \int x + \int y,\qquad(1)$$

but in the theory of algebraic integration, we have to be somewhat cautious. It is quite possible for $\int x + y$ to have an explicit form, even though $\int f$ and $\int g$ have not. One example of this phenomen is $\int x^x + (\log x)x^x$, whose integral is x^x, whilst its two summands do not have any integrals (in finite form). In fact, (1) must only be used if it is known that two of the three integrals exist.

A polynomial p always has a finite integral, so (1) holds for $x = p$ and $y = q/r$. Therefore the problem of integrating f reduces to the problem of the integration of p (which is very simple) and of the *proper* rational function q/r. The remainder of this section is devoted to the integration of a proper rational function.

5.1.2.1 *The naïve method*

If the polynomial r factorises into linear factors, such that

$$r = \prod_{i=1}^{n}(x - a_i)^{n_i},$$

we can decompose q/r into partial fractions (see the appendix "Algebraic background"):

$$\frac{q}{r} = \sum_{i=1}^{n}\frac{b_i}{(x - a_i)^{n_i}},$$

where the b_i are polynomials of degree less than n_i. These polynomials can be divided by $x - a_i$, so as to give the following decomposition:

$$\frac{q}{r} = \sum_{i=1}^{n}\sum_{j=1}^{n_i}\frac{b_{i,j}}{(x - a_i)^{j}},$$

where the $b_{i,j}$ are constants.

This decomposition can be integrated, then

$$\int \frac{q}{r} = \sum_{i=1}^{n} b_{i,j} \log (x - a_i) - \sum_{i=1}^{n} \sum_{j=2}^{n_i} \frac{b_{i,j}}{(j-1)(x - a_i)^{j-1}}.$$

Thus, we have proved that every rational function has an integral which can be expressed as a rational function plus a sum of logarithms of rational functions with constant coefficients — that is, that the integral belongs to the field $C(x, \log)$.

Although this proof is quite well known, it has several defects from the algorithmic point of view.

(1) It requires us to factorise r completely, which is not always possible without adding several algebraic quantities to C. As we have already seen, manipulating these algebraic extensions is often very difficult.

(2) Even if the algebraic extensions are not required, it is quite expensive to factorise a polynomial r of high degree.

(3) It also requires a big decomposition into partial fractions.

To prove that there are several integrals which can be determined without too much difficulty (with pencil and paper), let us consider

$$\int \frac{5x^4 + 60x^3 + 255x^2 + 450x + 274}{x^5 + 15x^4 + 85x^3 + 225x^2 + 274x + 120}$$
$$= \log(x^5 + 15x^4 + 85x^3 + 225x^2 + 274x + 120)$$
$$= \log(x + 1) + \log(x + 2) + \log(x + 3) + \log(x + 4) + \log(x + 5);$$

$$\int \frac{5x^4 + 60x^3 + 255x^2 + 450x + 275}{x^5 + 15x^4 + 85x^3 + 225x^2 + 274x + 120}$$
$$= \frac{25}{24} \log(x+1) + \frac{5}{6} \log(x+2) + \frac{5}{4} \log(x+3) + \frac{5}{6} \log(x+4) + \frac{25}{24} \log(x+5)$$

(the expression with a single logarithm is too long to write here: it contains the logarithm of a polynomial of degree 120, of which the largest coefficients have 68 decimal places);

$$\int \frac{5x^4 + 1}{(x^5 + x + 1)^2} = \frac{-1}{(x^5 + x + 1)};$$

$$\int \frac{5x^4 + 1}{(x^5 + x + 1)} = \log (x^5 + x + 1).$$

The first equation is an example where we can factorise the denominator (which is only $(x + 1)(x + 2)(x + 3)(x + 4)(x + 5)$), but the decomposition into partial fractions would be quite tiresome. The second one shows that a very small difference can greatly change the integral. The last two examples have denominators which do not factorise at all over $\mathbf{Q}$, and we have to add four algebraic extensions before it factorises completely (or an extension of degree 120!). So, we can state the following problem: *Find an algorithm of integration of rational functions which brings in only those algebraic quantities necessary for expressing the integral.*

5.1.2.2 *Hermite's method*

This method [Hermite, 1872] enables us to determine the rational part of the integral of a rational function without bringing in any algebraic quantity. Similarly, it finds the derivative of the sum of logarithms, which is also a rational function with coefficients in the same field as f. We have seen that a factor of the denominator r which appears to the power n, appears to the power $n - 1$ in the denominator of the integral. This suggests square-free decomposition (see the appendix "Algebraic background").

Let us suppose, then, that r has a square-free decomposition of the form $\prod_{i=1}^{n} r_i^i$. The r_i are then relatively prime, and we can construct a decomposition into partial fractions (see the appendix "Algebraic background"):

$$\frac{q}{r} = \frac{q}{\prod_{i=1}^{n} r_i^i} = \sum_{i=1}^{n} \frac{q_i}{r_i^i}.$$

We know that every element on the right hand side has an integral, and therefore (1) holds, and it suffices to integrate each element in turn. This is done with Bézout's identity (see once more the appendix "Algebraic background") applied to r_i and r_i', which are relatively prime, and hence there are an a and a b satisfying $ar_i + br_i' = 1$.

$$\int \frac{q_i}{r_i^i} = \int \frac{q_i(ar_i + br_i')}{r_i^i}$$

$$= \int \frac{q_i a}{r_i^{i-1}} + \int \frac{q_i br_i'}{r_i^i}$$

$$= \int \frac{q_i a}{r_i^{i-1}} + \int \frac{(q_i b/(i-1))'}{r_i^{i-1}} - \left(\frac{q_i b/(i-1)}{r_i^{i-1}}\right)'$$

$$= -\left(\frac{q_i b/(i-1)}{r_i^{i-1}}\right) + \int \frac{q_i a + (q_i b/(i-1))'}{r_i^{i-1}},$$

and we have been able to reduce the exponent of r_i. We can continue in this way until the exponent becomes one, when the remaining integral is a sum of logarithms.

5.1.2.3 *Horowitz' method*

Hermite's method is quite suitable for manual calculations, but the disadvantage of it is that it needs several sub-algorithms (square-free decomposition, decomposition into partial fractions, Bézout's identity), and this involves some fairly complicated programming. Therefore Horowitz [1969, 1971] proposed the following method.

The aim is still to be able to write $\int q/r$ in the form $q_1/r_1 + \int q_2/r_2$, where the integral remaining gives only a sum of logarithms when it is resolved. By Hermite's method and the discussion above, we know that r_1 has the same factors as r, but with the exponent reduced by one, and that r_2 has no multiple factors, and that its factors are all factors of r. The arguments of the section "Square-free decompositions" (in the appendix "Algebraic background") imply that $r_1 = \gcd(r, r')$, and that r_2 divides $r/\gcd(r, r')$. We may suppose that q_2/r_2 is written in reduced form, and therefore that $r_2 = r/\gcd(r, r')$.

Then

$$\frac{q}{r} = \left(\frac{q_1}{r_1}\right)' + \frac{q_2}{r_2}$$

$$= \frac{q_1'}{r_1} - \frac{q_1 r_1'}{r_1^2} + \frac{q_2}{r_2}$$

$$= \frac{q_1' r_2 - q_1 s + q_2 r_1}{r},$$

where $s = r_1' r_2/r_1$ (the division here is without remainder).

Thus, our problem reduces to a purely polynomial problem, that is

$$q = q_1' r_2 - q_1 s + q_2 r_1, \tag{2}$$

where q, s, r_1 and r_2 are known, and q_1 and q_2 have to be determined. But the degrees of q_1 and q_2 are less than the degrees m and n of r_1 and r_2 respectively. So, we write $q_1 = \sum_{i=0}^{m-1} a_i x^i$ and $q_2 = \sum_{i=0}^{n-1} b_i x^i$. Equation (2) is rewritten as a system of $m + n$ linear equations in $n + m$ unknowns. Moreover, this system can be solved, and integration (at least this subproblem) reduces to linear algebra.

5.1.2.4 *The logarithmic part*

The two methods described above can reduce the integration of any rational function to the integration of a rational function (say q/r) whose integral would be only a sum of logarithms. We know that this integral can be resolved by completely factorising the denominator, but, as we have already seen, this is not always necessary for an expression of the results. The real problem is to find the integral without using any algebraic numbers other than those needed in the expression of the result. This problem has been solved by Rothstein [1976] and by Trager [1976] — we give the latter's solution.

Let us suppose that

$$\int \frac{q}{r} = \sum_{i=1}^{n} c_i \log v_i \tag{3}$$

is a solution to this integral where the right hand side uses the fewest possible algebraic extensions. The c_i are constants and, in general, the v_i are rational functions. Since $\log(a/b) = \log a - \log b$, we can suppose, without loss of generality, that the v_i are polynomials. Furthermore, we can perform a square-free decomposition, which does not add any algebraic extensions, and we can apply the rule $\log \prod p_i^i = \sum i \log p_i$. Moreover, since $c \log pq + d \log pr = (c+d) \log p + c \log q + d \log r$, we can suppose that the v_i are relatively prime, whilst still keeping the minimality of the number of algebraic extensions (even though the number of logarithms may change). Moreover, we can suppose that all the c_i are different.

Differentiating equation (3), we find

$$\frac{q}{r} = \sum_{i=1}^{n} \frac{c_i v_i'}{v_i}. \tag{3'}$$

The assumption that the v_i are square-free implies that no element of this summation can simplify, and the assumption that the v_i are relatively prime implies that no cancellation can take place in this summation. This implies that the v_i must be precisely the factors of r, i.e. that $r = \prod_{i=1}^{n} v_i$. Let us write $u_i = \prod_{j \neq i} v_j$. Then we can differentiate the product of the v_i, which shows that $r' = \sum v_i' u_i$. If we clear the denominators in (3'), we find that $q = \sum c_i v_i' u_i$. These two expressions for q and r' permit the following deductions.

$$v_k = \gcd(0, v_k)$$
$$= \gcd\left(q - \sum c_i v_i' u_i, v_k\right)$$
$$= \gcd\left(q - c_k v_k' u_k, v_k\right)$$

since all the other u_i are divisible by v_k

$$= \gcd\left(q - c_k \sum v_i' u_i, v_k\right)$$

for the same reason

$$= \gcd\left(q - c_k r', v_k\right).$$

If $l \neq k$, we find that

$$\gcd\left(q - c_k r', v_l\right) = \gcd\left(\sum c_i v_i' u_i - c_k \sum v_i' u_i, v_l\right)$$

$$= \gcd\left(c_l v_l' u_l - c_k v_l' u_l, v_l\right)$$

since all the other u_i are divisible by v_l

$$= 1$$

since v_l has no repeated factors, or any factors in common with the product u_l of the other v_i. With the help of these calculations, we can proceed to the following calculation:

$$\gcd\left(q - c_k r', r\right) = \gcd\left(q - c_k r', \prod_{i=1}^{n} v_i\right)$$

$$= \prod_{i=1}^{n} \gcd\left(q - c_k r', v_i\right)$$

since the v_i are relatively prime

$$= \gcd\left(q - c_k r', v_k\right)$$

since all the other terms disappear

$$= v_k.$$

Hence, if we know the c_k, we can calculate the v_k. Furthermore, the c_k are precisely the values of y for which $\gcd(q - yr', r) \neq 1$. But these values can be calculated with the help of the resultant (see the appendix "Algebraic background"). $\mathrm{Res}_x(q - yr', r)$ is a polynomial in y, which is zero if, and only if, the g.c.d. is non-trivial. Hence it suffices to calculate this polynomial (which does not need any algebraic extensions) and, for each of its roots c_k, to determine $v_k = \gcd(q - c_k r', r)$.

5.1.3 The integration of more complicated functions

As soon as we leave the case of rational functions, we find functions whose integrals cannot be expressed in simpler terms. For example, we can say that

$$\int e^{-x^2} = \sqrt{\frac{\pi}{2}} \mathrm{erf} x,$$

but this is only a re-expression of the definition of the function erf. Now, one could remark that $\int 1/x = \log x$ is only a re-expression of the definition of the function log, and that is a valid comment. But the difference is that log is "well known" (a term that will be made precise later). In other words, which classes of functions B will we allow as classes of possible integrals. Liouville proposed the following definition, which seems quite suitable.

Definition. *Let K be a field of functions. The function θ is an elementary generator over K if:*

(a) *θ is algebraic over K, i.e. θ satisfies a polynomial equation with coefficients in K;*

(b) *θ is an exponential over K, i.e. there is an η in K such that $\theta' = \eta'\theta$, which is only an algebraic way of saying that $\theta = \exp \eta$;*

(c) *θ is a logarithm over K, i.e. there is an η in K such that $\theta' = \eta'/\eta$, which is only an algebraic way of saying that $\theta = \log \eta$.*

Definition. *Let K be a field of functions. An overfield $K(\theta_1, \ldots, \theta_n)$ of K is called a field of elementary functions over K if every θ_i is an elementary generator over K. A function is elementary over K if it belongs to a field of elementary functions over K.*

If K is omitted, we understand $\mathbf{C}(x)$: the field of rational functions.

This definition also includes the trigonometric functions — for example $\sin x$ is elementary (over $\mathbf{C}(x)$), since

$$\sin x = \frac{1}{2i}\left(e^{ix} + e^{-ix}\right) = \frac{1}{2i}\left(\theta + \frac{1}{\theta}\right),$$

where θ is the exponential of ix. This is also true for the inverses of trigonometric functions — for example $\tan^{-1} x = \log\left(\frac{x+i}{x-i}\right)$. We will use the notation $K(\text{elem})$ for the class of elementary functions over K.

The previous section showed that every rational function has an elementary integral. Furthermore, this integral has a particular form — a rational function (whose calculation needs no algebraic extensions) plus a sum of logarithms with constant coefficients. We shall find that this form holds in a much more general context, as the following result shows.

Theorem (Liouville's Principle). *Let f be a function from some function field K. If f has an elementary integral over K, it has an integral of the following form:*

$$\int f = v_0 + \sum_{i=1}^{n} c_i \log v_i, \tag{4}$$

where v_0 belongs to K, the v_i belong to $\hat{K}$, an extension of K by a finite number of constants algebraic over K, and the c_i belong to $\hat{K}$ and are constant.

Another way of putting this is to say that, if f has an elementary integral over K, then f has the following form:

$$f = v_0' + \sum_{i=1}^{n} \frac{c_i v_i'}{v_i}. \tag{4'}$$

This theorem is fundamental for the theory of integration in terms of elementary functions. In what follows, we shall use the expression *"We can integrate in A with results in B"* as an abbreviation of "There exists an algorithm which, given an element a of A, returns an element b of B such that $a = b'$, or shows that no function of B has this property". If we omit B, we mean the class A(elem).

5.1.4 Integration of logarithmic functions

This section is devoted to the study of fields of functions generated by a single *logarithm*. We shall work in the following setting:
  (i) K is a field of functions, assumed to be effective, such that we can integrate in K;
 (ii) θ is a logarithm over K, $\theta' = \eta'/\eta$, and θ is assumed transcendental over K;
(iii) $K(\theta)$ has the same constants as K.
This last hypothesis avoids the difficulties which can occur if θ gives rise to new constants. For example, if $K = \mathbf{Q}(x, \log x)$ and $\theta = \log 2x$, $K(\theta)$ contains the constant $\log 2$. This hypothesis, as well as the hypothesis that θ is transcendental, can be checked by means of the Risch Structure Theorem (see the section "Representation of Transcendental Functions"). The aim is to show that we can integrate in $K(\theta)$. On the whole, the methods are similar to those used in the integration of rational functions, but there are several occasions when a little care is necessary.

5.1.4.1 *The decomposition lemma*

We can write our function f of $K(\theta)$ in the form $p + q/r$, where p, q and r are polynomials of $K[\theta]$, q and r are relatively prime, and the degree of q is less than that of r. Can we, then, treat p and q/r separately, i.e. can we apply the decomposition (1)?

Decomposition Lemma. *If f has an elementary integral over K, then p and q/r each possess an elementary integral over K.*

Proof. From Liouville's principle (in the form of equation (4')),

$$p + \frac{q}{r} = v_0' + \sum_{i=1}^{n} \frac{c_i v_i'}{v_i}.$$

In this decomposition, we can suppose that $v_1, \ldots, v_k$ are monic polynomials in θ whose coefficients belong to K, whilst $v_{k+1}, \ldots, v_n$ belong to K. Furthermore, we can write v_0 in the form $\hat{p} + \hat{q}/\hat{r}$, where $\hat{p}$, $\hat{q}$ and $\hat{r}$ are polynomials in θ, and $\hat{q}/\hat{r}$ is a proper rational function. With these conventions and using the following properties of differentiation —

(a) the derivative of a polynomial is a polynomial;
(b) the derivative of a proper rational function is itself a proper rational function —

we see that

$$p + \frac{q}{r} = \left(\hat{p} + \frac{\hat{q}}{\hat{r}}\right)' + \sum_{i=1}^{n} \frac{c_i v_i'}{v_i}$$

$$= \underbrace{\hat{p}' + \sum_{i=k+1}^{n} \frac{c_i v_i'}{v_i}}_{\text{polynomial}}$$

$$+ \underbrace{\left(\frac{\hat{q}}{\hat{r}}\right)' + \sum_{i=1}^{k} \frac{c_i v_i'}{v_i}}_{\text{proper rational function}}$$

Since the decomposition "polynomial + proper rational function" is unique (remember that θ is assumed to be transcendental), this equation lets us deduce the following two equations:

$$p = \hat{p}' + \sum_{i=k+1}^{n} \frac{c_i v_i'}{v_i};$$

$$\frac{q}{r} = \left(\frac{\hat{q}}{\hat{r}}\right)' + \sum_{i=1}^{k} \frac{c_i v_i'}{v_i}.$$

These two equations can be integrated formally, in order to give

$$\int p = \hat{p} + \sum_{i=k+1}^{n} c_i \log v_i;$$

$$\int \frac{q}{r} = \frac{\hat{q}}{\hat{r}} + \sum_{i=1}^{k} c_i \log v_i.$$

These equations prove the stated result, and, furthermore, they provide further information about the shape of the integrals. In the following subsections, we consider these two parts separately.

5.1.4.2 *The polynomial part*

Suppose p is written in the form $\sum_{i=0}^{m} A_i \theta^i$, and that $\hat{p}$ is written in the form $\sum_{i=0}^{n} B_i \theta^i$. The last equations in the proof of the decomposition lemma imply that

$$\sum_{i=0}^{m} A_i \theta^i = \left(\sum_{i=0}^{n} B_i \theta^i\right)' + \sum_{i=k+1}^{n} \frac{c_i v_i'}{v_i}$$

$$= \sum_{i=0}^{n} B_i' \theta^i + \sum_{i=1}^{n} i B_i \theta' \theta^{i-1} + \sum_{i=k+1}^{n} \frac{c_i v_i'}{v_i}.$$

We recall that the v_i which occur in this expression belong to K. Since θ is transcendental, we can deduce that the coefficients of each power of θ in the two parts must be equal. This implies that $n = m$ or $n = m + 1$. The coefficients of θ^{m+1} give the equation $0 = B_{m+1}'$, from which we deduce that B_{m+1} must be a constant.

The coefficients of θ^m (supposing that $m > 0$) give the equation:

$$A_m = B_m' + (m+1)B_{m+1}\theta'.$$

This equation can be rewritten in an integral form:

$$B_m = \int \left(A_m - (m+1)B_{m+1}\theta'\right)$$

$$= -(m+1)B_{m+1}\theta + \int A_m.$$

This equation implies that A_m must have an elementary integral, and hypothesis (i) of this section tells us that there is an algorithm for finding this integral (or for proving that this function does not have an elementary integral, which means that p does not have one either). Moreover, this integral must be of a special form: an element of K plus a (constant) multiple of θ. θ is the only logarithm which may be added to K in order to express this integral. (It is therefore possible to optimise this integration of A_m but this makes the presentation much more complicated, and so we shall not do it here.) This integration determines B_{m+1}, and determines B_m to within a constant. Let us call this constant b_m.

The coefficients of θ^{m-1} (supposing $m > 1$) give the equation:

$$A_{m-1} = B_{m-1}' + m(B_m + b_m)\theta'.$$

This equation can be rewritten in an integral form:

$$B_{m-1} = \int \left(A_{m-1} - m(B_m + b_m)\theta'\right)$$

$$= -mb_m\theta + \int A_{m-1} - mB_m\theta'. \tag{5}$$

As we have already done for the coefficients of θ^m, we can deduce that $A_{m-1} - mB_m\theta'$ has an elementary integral, which must belong to K, except possibly for a constant multiple of θ, which determines b_m. Thus, B_{m-1} is determined to within a constant, which we shall call b_{m-1}.

We continue in this way as far as the coefficients of θ^0:

$$A_0 = B_0' + 1(B_1 + b_1)\theta' + \sum_{i=k+1}^{n} \frac{c_i v_i'}{v_i}.$$

This equation can be rewritten in an integral form:

$$B_0 + \sum_{i=k+1}^{n} c_i \log v_i = \int (A_0 - (B_1 + b_1)\theta')$$

$$= -b_1\theta + \int A_0 - B_1\theta'.$$

We deduce from this that $A_0 - B_1\theta'$ has an elementary integral. This time, it is not necessary for the integral to belong to K, and it can contain new logarithms. The coefficient of θ in this integral gives us b_1. As before, B_0 is only determined to within a constant, but this constant is the constant of integration, which remains undetermined. Thus we have found the whole of the integral of the polynomial part (or else we have shown that there is none).

5.1.4.3 *The rational and logarithmic part*

The last equations in the proof of the decomposition lemma imply that

$$\frac{q}{r} = \left(\frac{\hat{q}}{\hat{r}}\right)' + \sum_{i=1}^{k} \frac{c_i v_i'}{v_i}.$$

The problem is similar to that of the rational functions, and the solution is also similar. In fact, Hermite's method works in this case as it does for rational functions, except that the meaning of the symbol $'$ must be taken into account. In the case of a polynomial in the variable x,

$$\left(\sum_{i=0}^{n} a_i x^i\right)' = \sum_{i=1}^{n} i a_i x^{i-1},$$

whilst for a polynomial in the variable θ,

$$\left(\sum_{i=0}^{n} a_i \theta^i\right)' = \sum_{i=1}^{n} i a_i \theta' \theta^{i-1} + \sum_{i=0}^{n} a_i' \theta^i.$$

Although they have to be proved again, all the necessary results in the appendix "Algebraic background" still hold, and Hermite's method works (or, more exactly, the Hermite-Ostrowski method, for it was Ostrowski [1946] who made this generalisation).

The logarithmic part is calculated in almost the same way as in the case of rational functions. We have proved that the coefficients of the logarithms, which by Liouville's theorem have to be constants, are the roots of the polynomial $\mathrm{Res}_\theta(q - yr', r)$. But in the present context, we can no longer be certain that these roots are constants. Let us take the example of $\int dx / \log x$, where $q = 1$ and $r = \log x$.

$$\mathrm{Res}_{\log x}\left(1 - y\frac{1}{x}, \log x\right) = \left|1 - \frac{y}{x}\right| = \frac{x - y}{x},$$

and the root of this polynomial is $y = x$, which is not a constant. In this case, we can conclude that the integral is not elementary, for an elementary integral must have constant coefficients. If we apply the same method to $\int dx / x \log x$, we find the polynomial $y = 1$, and the integral becomes $\log \log x$.

5.1.5 Integration of exponential functions

This section is devoted to the study of functions generated by an *exponential*. Here we have the same process as in the last section, except where the differences in behaviour between logarithmic and exponential functions compel us to make distinctions. In this section, we are operating within the following context:

  (i) K is a field of functions, supposed effective, such that we know how to integrate in K and such that we know how to solve the differential equation $y' + fy = g$ in K (see the next section for the solution of this problem);

 (ii) θ is an exponential over K, $\theta' = \eta'\theta$, and θ is supposed transcendental over K;

(iii) $K(\theta)$ has the same constants as K.

The last hypothesis avoids the difficulties which may arise if θ gives rise to new constants. For example, if $K = \mathbf{Q}(x, \exp x)$ and $\theta = \exp(x + 1)$, $K(\theta)$ contains the constant e. This hypothesis, like the hypothesis that θ is transcendental, can be verified using the Risch Structure Theorem (see the section "Representation of transcendental functions"). The aim is to prove that we know how to integrate in $K(\theta)$. The methods for the rational part are to a large extent similar to those for the integration of logarithmic functions, but the polynomial part is rather different.

We must point out that the choice of θ is somewhat arbitrary. $\frac{1}{\theta}$ is also an exponential: that of $-\eta$. So it is obvious that the "polynomial part"

must take this symmetry into account. In the last section, we used the
ideas of square-free decomposition etc., saying they hold also for the case
of polynomials in one logarithmic variable. But they do not hold in the
case of an exponential variable: for example, if $\theta = \exp x$, the polynomial
$p = \theta$ has no multiple factors, but $\gcd(p, p') = \theta$, and such a non-trivial
g.c.d. normally implies that p has a multiple factor. It is not difficult to
verify that these concepts are valid if θ does not divide p.

5.1.5.1 *The decomposition lemma*

We can write this function f of $K(\theta)$ in the form $p + q/r$, where p is a
generalised polynomial (that is $\sum_{i=-m}^{n} a_i \theta^i$), q and r are polynomials of
$K[\theta]$ such that θ does not divide r, q and r are relatively prime, and the
degree of q is less than that of r. So can we deal with p and q/r separately,
that is, apply the decomposition (1)?

Decomposition lemma. *If f has an elementary integral over K, then
each of the terms of p, and also q/r, have an elementary integral over K.*

Proof. By Liouville's principle (in the form of equation (4')),

$$p + \frac{q}{r} = v_0' + \sum_{i=1}^{n} \frac{c_i v_i'}{v_i}.$$

In this decompositon, we can suppose that $v_1, \ldots, v_k$ are monic polynomials
in θ of degree n_i (the coefficients of which belong to K), whereas $v_{k+1}, \ldots, v_n$
belong to K. We can also suppose that none of the v_i is divisible by θ, for
$\log \theta = \eta$. Moreover, we can write v_0 in the form $\hat{p} + \hat{q}/\hat{r}$, in the same
way as we broke down f. With these conventions and using the properties
that the derivative of a generalised polynomial is a generalised polynomial,
and that the derivative of a proper rational function (such that θ does not
divide its denominator) is a proper rational function (such that θ does not
divide its denominator), we see that:

$$p + \frac{q}{r} = \left(\hat{p} + \frac{\hat{q}}{\hat{r}}\right)' + \sum_{i=1}^{n} \frac{c_i v_i'}{v_i}$$

$$= \hat{p}' + \underbrace{\sum_{i=k+1}^{n} \frac{c_i v_i'}{v_i}}_{\text{part "p"}} + \underbrace{\left(\frac{\hat{q}}{\hat{r}}\right)' + \sum_{i=1}^{k} \frac{c_i v_i'}{v_i}}_{\text{part "r"}}.$$

In the section on logarithmic functions, we remarked that the two parts of
this decomposition were a polynomial and a proper rational function. This

observation does not hold for the present case. Let us suppose that v_i (with $i \leq k$) is written in the form $\theta^{n_i} + \sum_{j=0}^{n_i-1} a_i \theta^i$. The polynomial v_i' is also a polynomial of degree n_i, that is

$$n_i \eta' \theta^{n_i} + \left(\sum_{j=0}^{n_i-1} a_i \theta^i \right)',$$

and therefore $c_i v_i'/v_i$ is not a proper rational fraction. However,

$$\frac{c_i v_i'}{v_i} - c_i n_i \eta' = \frac{c_i(v_i' - n_i \eta' v_i)}{v_i}$$

is a proper rational fraction. We therefore have the following decomposition:

$$p + \frac{q}{r} = \hat{p}' + \underbrace{\sum_{i=k+1}^{n} \frac{c_i v_i'}{v_i} + \eta' \sum_{i=1}^{k} c_i n_i}_{\text{generalised polynomial}}$$

$$+ \underbrace{\left(\frac{\hat{q}}{\hat{r}} \right)' + \sum_{i=1}^{k} \frac{c_i(v_i' - n_i \eta' v_i)}{v_i}}_{\text{proper rational function}}$$

(where, moreover, θ does not divide the denominator of the second part).

Since decompositions of the form "generalised polynomial + proper rational function" (of which θ does not divide the denominator) are unambiguous (we recall that θ is supposed transcendental), we can deduce from this equation the following two equations:

$$p = \hat{p}' + \sum_{i=k+1}^{n} \frac{c_i v_i'}{v_i} + \eta' \sum_{i=1}^{k} c_i n_i;$$

$$\frac{q}{r} = \left(\frac{\hat{q}}{\hat{r}} \right)' + \sum_{i=1}^{k} \frac{c_i(v_i' - n_i \eta' v_i)}{v_i}.$$

The second equation integrates formally to give:

$$\int \frac{q}{r} = \frac{\hat{q}}{\hat{r}} + \sum_{i=1}^{k} c_i(\log v_i - n_i \eta).$$

In the first equation we write $p = \sum_{i=-m}^{n} A_i \theta^i$ and $\hat{p} = \sum_{i=-m'}^{n'} B_i \theta^i$, where the A_i and B_i belong to K. Since θ is transcendental, we can deduce that the coefficients of each power of θ in the two parts must be equal. This implies that $n = n'$ and $m = m'$. Since the derivative of $B_i\theta^i$ is $(B_i' + i\eta'B_i)\theta^i$, we find that, for $i \neq 0$, $A_i\theta^i = (B_i' + i\eta'B_i)\theta^i$, which implies that all these latter terms have integrals. Since the rational part also has an integral, the remaining term, that is A_0, must also have an integral.

5.1.5.2 *The generalised polynomial part*

We have just proved that, for $i \neq 0$,

$$\int A_i\theta^i = B_i\theta^i,$$

where B_i belongs to k and satisfies the differential equation $B_i' + i\eta'B_i = A_i$. By hypothesis, we have an algorithm which solves this problem in K, either determining B_i, or proving that no B_i in K satisfies this equation (see the section "Ordinary differential equations in finite form" for the details).

A_0 belongs to K, and therefore its integral can be determined by the algorithm of integration in K. It must be noted that we have to subtract $\eta\sum_{i=1}^{k} c_i n_i$ from this integral to find the integral of f, for this sum was added in the process of decomposition.

5.1.5.3 *The rational and logarithmic part*

The last equations in the proof of the decomposition lemma imply that

$$\frac{q}{r} = \left(\frac{\hat{q}}{\hat{r}}\right)' + \sum_{i=1}^{k} \frac{c_i(v_i' - n_i\eta'v_i)}{v_i}.$$

In fact, Hermite's method holds in this case as it does for rational and logarithmic functions, since θ does not divide the denominator r, and therefore the square-free decomposition $r = \prod_{i=1}^{n} r_i^i$ works, and moreover each r_i and its derivative are relatively prime.

We can therefore return to the case where r has no multiple factors. Previously, that indicated that the integral was a sum of logarithms, but here it implies that

$$\int \frac{q}{r} = \sum_{i=1}^{k} c_i(\log v_i - n_i\eta).$$

The method we shall use for finding the c_i and the v_i is similar to that used for rational functions of x, but it requires some technical modifications. As before, we can suppose that the v_i are square-free polynomials, and relatively prime. The analogue of equation (3') is

$$\frac{q}{r} = \sum_{i=1}^{n} \frac{c_i(v_i' - n_i\eta'v_i)}{v_i}. \tag{6}$$

There cannot be any cancellation in this sum, and this means that the v_i must be precisely the factors of r, that is $r = \prod_{i=1}^{n} v_i$. We write $u_i = \prod_{j \neq i} v_j$. Then we can differentiate the product of the v_i, in order to determine that $r' = \sum v_i' u_i$. If we clear denominators in (6), we find $q = \sum c_i (v_i' - n_i \eta' v_i) u_i$.

$$v_k = \gcd(0, v_k)$$
$$= \gcd\left(q - \sum c_i(v_i' - n_i\eta'v_i)u_i, v_k\right)$$
$$= \gcd\left(q - c_k(v_k' - n_k\eta'v_k)u_k, v_k\right)$$

for all the other u_i are divisible by v_k

$$= \gcd\left(q - c_k \sum (v_i' - n_i\eta'v_i)u_i, v_k\right)$$

for the same reason

$$= \gcd\left(q - c_k(r' - \eta'r\sum n_i), v_k\right).$$

If $l \neq k$, we find, as we do for rational functions,

$$\gcd\left(q - c_k(r' - \eta'r\sum n_i), v_l\right) = 1.$$

With the help of these two calculations, we determine

$$\gcd\left(q - c_k(r' - \eta'r\sum n_i), r\right) = v_k.$$

This formula uses $\sum n_i$, but this is only the degree of r, which we can call N. Thus, if we know c_k, we can calculate v_k. Moreover, the c_k are precisely those values of y for which $\gcd(q - y(r' - N\eta'r), r) \neq 1$. But these values can be calculated using the concept of the resultant (see the appendix "Algebraic background"). $\text{Res}_\theta(q - y(r' - N\eta'r), r)$ is a polynomial in y, which cancels if and only if the g.c.d. is non-trivial. Therefore it suffices to calculate this polynomial (which can be done without any algebraic extension), to find its roots (which may indeed require some algebraic extensions), and, for each of its roots c_k, to determine $v_k = \gcd(q - c_k(r' - N\eta'r), r)$. As in the case of logarithmic functions, it is possible that this polynomial has some non-constant roots, which would imply that f has no elementary integral.

5.1.6 Integration of mixed functions

It is possible that there are functions which are not purely logarithmic or purely exponential. But the hypotheses on K we have stated enable us to integrate mixed functions, by considering the function as a member of

$K(\theta)$, where K is a field of functions and θ is a logarithm or an exponential. For example, let us consider the function

$$\frac{-e^x \log^2 x + \log x \left(\dfrac{2(e^x + 1)}{x}\right) + e^x + e^{2x}}{1 + 2e^x + e^{2x}}.$$

The problem. This function belongs to the field of functions $Q(x, e^x, \log x)$. Therefore we can write $K = Q(x, e^x)$, $\theta = \log x$ and apply the theory of the section "Integration of logarithmic functions". As an element of $K(\theta)$, this function is a polynomial in θ:

$$\theta^2 \left(\frac{-e^x}{1 + 2e^x + e^{2x}}\right) + \theta \left(\frac{2}{x(1 + e^x)}\right) + \frac{e^x}{1 + e^x}.$$

Following the method of the section "Integration of logarithmic functions — the polynomial part", we must integrate the coefficient of the leading power of θ, that is $-e^x/(1 + e^x)^2$. This integration takes place in K.

Sub-problem 1. This integration takes place in the field $L(\phi)$, where $\phi = e^x$ and $L = Q(x)$. The function to be integrated is a proper rational function, and, moreover, ϕ does not divide the denominator. Therefore the theory of the section "Integration of exponential functions — the rational and logarithmic part" applies. Square-free decomposition is quite easy, and we only have to apply Hermite's method to q/r^2 where $q = -\phi$ and $r = 1 + \phi$. We find that $r' = \phi$ (the symbol $'$ always denotes differentiation with respect to x). The Bézout identity has to be applied to r and r', which is quite easy in the present case:

$$(1)r + (-1)r' = 1.$$

By substituting these values in Hermite's method, we find an integral of $-q(-1)/(1 + \phi)$, and a remainder which cancels completely. Thus the solution of this sub-problem is $-\phi/(1 + \phi)$.

In the original problem, this gives us a term of the integral, that is $-\theta^2 e^x/(1 + e^x)$. But the derivative of this term gives us also terms in θ^1. The coefficient of θ^1 to be integrated, according to formula (5) of the section "The polynomial part", is given by $A_1 - 2\theta' B_2$, where A_1 is the original coefficient and B_2 is the solution of the sub-problem 1. This calculation gives

$$\frac{2}{x(1 + e^x)} - \frac{2}{x} \frac{-e^x}{1 + e^x} = \frac{2}{x}.$$

The integral of this function ought to give the coefficient of θ in the integral (and, possibly, determine the constant of integration in the previous sub-problem).

Sub-problem 2. In theory this integration takes place in the field $L(\phi)$, where $\phi = e^x$ and $L = \mathbf{Q}(x)$, but in reality it is quite easy to see that the answer is $2 \log x$.

In general, the integrals given by the sub-problems must belong to K, but, as we have seen, they may be allowed to contain θ. That is the case here, for the integral is 2θ. This implies that the choice of the constant of integration in the last sub-problem was bad, and that it must be increased by $2/2 = 1$. So, the present state of this problem is that we have integrated the coefficients of θ^2 and of θ^1, and we have found the integral to be $\theta^2(1 - e^x/(1 + e^x))$, which simplifies into $\theta^2/(1 + e^x)$. We still have to integrate the coefficient of θ^0, that is $e^x/(1 + e^x)$.

Sub-problem 3. This integration takes place in the field $L(\phi)$, where $\phi = e^x$ and $L = \mathbf{Q}(x)$. The function to be integrated is not a proper rational function, and has to be rewritten in the form $1 - 1/(1 + e^x)$. Integration of 1 gives x (plus a constant of integration, which is the integration constant of the problem). The other part is a proper rational function q/r, where $q = -1$ and $r = 1 + e^x$. r is not divisible by e^x and has no multiple factors, therefore its integral must be a sum of logarithms. By the theory of the section "Integration of exponential functions — the rational and logarithmic part", we have to calculate

$$\text{Res}_\phi(q - y(r' - N\eta'r), r),$$

where $N = 1$ (the degree of r), and $\eta = x$. This simplifies into $\text{Res}_\phi(-1 + y, 1 + \phi)$, that is $-1 + y$. This polynomial has a root, $y = 1$, which is indeed a constant. Therefore the integral of the rational part is $\log(1 + e^x) - x$. The $-x$ cancels with the x of the other part, and we have $\log(1 + e^x)$.

Thus the solution of the problem is

$$\log^2 x \left(\frac{1}{1 + e^x} \right) + \log(1 + e^x).$$

This apparently rather complicated method for breaking down an integral into a series of nested, but simple, problems is actually very general. It gives the following result.

Theorem [Risch, 1969]. *Let $K = C(x, \theta_1, \theta_2, \ldots, \theta_n)$ be a field of functions, where C is a field of constants and each θ_i is a logarithm or an exponential of an element of $C(x, \theta_1, \theta_2, \ldots, \theta_{i-1})$, and is transcendental over $C(x, \theta_1, \theta_2, \ldots, \theta_{i-1})$. Moreover, the field of constants of K must be C. Then there is an algorithm which, given an element f of K, either gives* *an elementary function over K which is the integral of f, or proves that f has no elementary integral over K.*

The proof is by induction on n ($n = 0$ being the integration of rational functions), using the theory of the sections "Integration of logarithmic functions" and "Integration of exponential functions". This theorem applies also to trigonometric functions (and their inverses), for they can be expressed as exponentials (or as logarithms).

5.1.7 Integration of algebraic functions

This is quite a difficult problem, which has interested many great mathematicians, and which is studied in "Algebraic geometry". After several advances in this subject, it was possible to give an answer [Davenport, 1981] to this problem, that is an algorithm which, given an algebraic function f over $\mathbf{Q}(x)$, either gives a function elementary over $\mathbf{Q}(x)$ which is its intgeral, or it proves that there is no such function.

This algorithm is fairly complicated and Davenport has only programmed that part of it where f is defined by square roots and where the geometrical problems are not too complicated*. More recently, Trager [1985] has given other methods which appear to be more efficient, but they have not yet been programmed. In fact, this problem is still at the frontier of our mathematical knowledge and of the power of our Computer Algebra systems. This algorithm can be used, instead of the integration of rational functions, as a starting point for the induction in Risch's theorem. This gives the following result.

Theorem [Davenport, 1984a]. *Let $K = C(x, y, \theta_1, \theta_2, \ldots, \theta_n)$ be a field of functions, where C is a field of constants, and y is algebraic over $C(x)$ and each θ_i is a logarithm or an exponential of an element of $C(x, y, \theta_1, \theta_2, \ldots, \theta_{i-1})$, and is transcendental over $C(x, y, \theta_1, \theta_2, \ldots, \theta_{i-1})$. Moreover, the field of constants of K must be C. Then there is an algorithm which, given an element f of K, either gives an elementary function over K which is the integral of f, or proves that f has no elementary integral over K.*

It is important to note that this theorem does not allow algebraic extensions which depend on logarithmic or exponential functions. This problem is still unsolved.

5.1.8 Integration of non-elementary functions

We have quoted Liouville's definition of the word "elementary", and have

---

* More precisely, though somewhat technically: the algebraic function must be defined over an algebraic curve of genus at most 1 if there are logarithmic parts to be calculated.

outlined a theory for integrating these functions, where the integrals too have to be elementary. What happens if we remove this restriction? Firstly, we can no longer use Liouville's principle. Recently Singer, Saunders and Caviness [1981, 1985] have generalised this principle, but the generalisation is quite complicated and only applies to a fairly restricted class of functions, even though this class is wider than the elementary functions. In fact, a finite number of functions of type E or L are allowed in addition to logarithms and exponentials.

Definition. *A function is of type E over a field K if it is of the form $\int u'G(\exp(R(u)))$, where $u \in K$ and G and R are rational functions with constant coefficients. A function is of type L over a field K if it is of the form $\int u'H(\log(S(u)))$, where $u \in K$ and H and S are rational functions with constant coefficients, such that the degree of the numerator of H is at most the degree of the denominator of H plus one.*

For example, the function $\operatorname{erf} x$ is of type E over $\mathbf{C}(x)$, with $u = x$ and the rational functions $G(y) = \sqrt{2/\pi}y$ and $R(y) = -y^2$. Similarly,

$$\operatorname{erf} \log x = \int (\sqrt{2/\pi}/x) exp(-\log^2 x)$$

is of type E over $\mathbf{C}(x, \log x)$. The function $\operatorname{Ei}(x) = \int e^x/x$ is not of type E over $\mathbf{C}(x)$, because the necessary function $G(y)$, say y/x, does not have constant coefficients. The function $\operatorname{Li}(x) = \int 1/\log x$ is of type L over $\mathbf{C}(x)$, with $u = x$, $H(y) = 1/y$ and $S(y) = y$. The function $\operatorname{Ei}(x)$ is of type L over $\mathbf{C}(x, e^x)$, since it can be written in the form $\operatorname{Li} e^x$.

To integrate in Computer Algebra requires more than Liouville's principle — it needs an algorithm too. There are now algorithms which generalise Risch's theorem by allowing only the same class of functions, but a larger class of integrals, that is functions which are "elementary with erf" or "elementary with Li (and Ei)" [Cherry, 1983, 1985; Cherry and Caviness, 1984]. But this is currently a subject of intensive study and new algorithms may be found.

5.2 ALGEBRAIC SOLUTIONS OF O.D.E.S

The problem of integration can be considered as solving the simplest differential equation: $y' = f$. In this section we state some facts about the behaviour of formal solutions of some linear differential equations. The remarks made in the previous section about the difference between formal solutions and numerical solutions still apply in this case, as do the remarks about the difference between the heuristic and the algorithmic approaches.

5.2.1 First order equations

Here we consider the equation $y' + fy = g$. The previous sub-section, on the integration of exponential functions, has already introduced the problem of finding an algorithm which, given f and g belonging to a class A of functions, either finds a function y belonging to a given class B of functions, or proves that there is no element of B which satisfies the given equation. For the sake of simplicity, we shall consider the case when B is always the class of functions elementary over A.

There is a fairly well-known method for solving equations of this kind: we substitute $y = ze^{-\int f}$. This gives

$$
\begin{aligned}
g &= y' + fy \\
&= z'e^{-\int f} - zfe^{-\int f} + fze^{-\int f} \\
&= z'e^{-\int f}.
\end{aligned}
$$

Thus $z' = ge^{\int f}$, and

$$
y = e^{-\int f} \int ge^{\int f}. \tag{1}
$$

In general, this method is not algorithmically satisfactory for finding y, since the algorithm of integration described in the last section reformulates this integral as the differential equation we started with.

5.2.1.1 *Risch's problem*

So we have to find a direct method for solving these equations. Risch [1969] found one for the case when A is a field of rational functions, or an extension (by a transcendental logarithm or by a transcendental exponential) of a field over which this problem can be solved. Here we give the algorithm for the case of rational functions: for the other cases the principles are similar but the details are much more complicated. The solution given here is, *grosso modo*, that of Risch [1969]: there is another one by Davenport [1985c].

The problem can be stated as follows: given two rational functions f and g, find the rational function y such that $y' + fy = g$, or prove that there is none. f satisfies the condition that $exp(\int f)$ is not a rational function, that is that f is not constant, and its integral is not a sum of logarithms with rational coefficients. The problem is solved in two stages: reducing it to a purely polynomial problem, and solving that problem.

Let p be an irreducible polynomial. Let α be the largest integer such that p^α divides the denominator of y, which we can write as $p^\alpha \parallel \mathrm{den}(y)$. Let β and γ be such that $p^\beta \parallel \mathrm{den}(f)$ and $p^\gamma \parallel \mathrm{den}(g)$. So we can calculate the powers of p which divide the terms of the equation to be solved:

$$\underbrace{y'}_{\alpha+1} + \underbrace{fy}_{\alpha+\beta} = \underbrace{g}_{\gamma}.$$

there are three possibilities.

(1) $\beta > 1$. In this case the terms in $p^{\alpha+\beta}$ and p^{γ} have to cancel, that is we must have $\alpha = \gamma - \beta$.

(2) $\beta < 1$ (in other words, $\beta = 0$). In this case the terms in $p^{\alpha+1}$ and p^{γ} must cancel, that is, we must have $\alpha = \gamma - 1$.

(3) $\beta = 1$. In this case, it is possible that the terms on the left-hand side cancel and that the power of p which divides the denominator of $y' + fy$ is less than $\alpha+1$. If there is no cancellation, the result is indeed $\alpha = \gamma - 1 = \gamma - \beta$. So let us suppose that there is a cancellation. We express f and y in partial fractions with respect to p: $f = F/p^{\beta} + \hat{f}$ and $y = Y/p^{\alpha} + \hat{y}$, where the powers of p which divide the denominators of $\hat{f}$ and $\hat{y}$ are at most $\beta - 1 = 0$ and $\alpha - 1$.

$$y' + fy = \frac{-\alpha p'Y}{p^{\alpha+1}} + \frac{Y'}{p^{\alpha}} + \frac{FY}{p^{\alpha+1}} + \frac{\hat{f}Y}{p^{\alpha}} + \frac{F\hat{y}}{p} + \hat{f}\hat{y}.$$

For there to be a cancellation in this equation, p must divide $-\alpha p'Y + FY$. But p is irreducible and Y is of degree less than that of p, therefore p and Y are relatively prime. This implies that p divides $\alpha p' - F$. But p' and F are of degree less than that of p, and the only polynomial of degree less than that of p and divisible by p is zero. Therefore $\alpha = F/p'$.

We have proved the following result:

Lemma [Risch, 1969]. $\alpha \le \max(\min(\gamma - 1, \gamma - \beta), F/p')$, where the last term only holds when $\beta = 1$, and when it gives rise to a positive integer.

In fact, it is not necessary to factorise the denominators into irreducible polynomials. It is enough to find square-free polynomials p_i, relatively prime in pairs, and non-negative integers β_i and γ_i such that $\mathrm{den}(f) = \prod p_i^{\beta_i}$ and $\mathrm{den}(g) = \prod p_i^{\gamma_i}$. When $\beta = 1$, we have, in theory, to factorise p completely, but it is enough to find the integral roots of $\mathrm{Res}_x(F - yp', p)$, by an argument similar to Trager's algorithm for calculating the logarithmic part of the integral of a rational function.

We have, therefore, been able to bound the denominator of y by $D = \prod p_i^{\alpha_i}$, so that $y = Y/D$ with Y polynomial. So it is possible to suppress the denominators in our equation, and to find an equation $RY' + SY = T$.

Let α, β, γ and δ be the degress of Y, R, S and T. There are three possibilities*.

(1) $\beta - 1 > \gamma$. In this case, the terms of degree $\alpha + \beta - 1$ must cancel out the terms of degree δ, therefore $\alpha = \delta + 1 - \beta$.

(2) $\beta - 1 < \gamma$. In this case, the terms of degree $\alpha + \gamma$ must cancel out the terms of degree δ, therefore $\alpha = \delta - \gamma$.

(3) $\beta - 1 = \gamma$. In this case, the terms of degree $\alpha + \beta - 1$ on the left may cancel. If not, the previous analysis still holds, and $\alpha = \delta + 1 - \beta$. To analyse the cancellation, we write $Y = \sum_{i=0}^{\alpha} y_i x^i$, $R = \sum_{i=0}^{\beta} r_i x^i$ and $S = \sum_{i=0}^{\gamma} s_i x^i$. The coefficients of the terms of degree $\alpha + \beta - 1$ are $\alpha r_\beta y_\alpha$ and $s_\gamma y_\alpha$. The cancellation is equivalent to $\alpha = -s_\gamma / r_\beta$.

We have proved the following result:

Lemma [Risch, 1969]. $\alpha \leq \max(\min(\delta - \gamma, \delta + 1 - \beta), -s_\gamma / r_\beta)$, where the last term is included only when $\beta = \gamma + 1$, and only when it gives rise to a positive integer.

Determining the coefficients y_i of Y is a problem of linear algebra. In fact, the system of equations is triangular, and is easily solved.

5.2.1.2 *A theorem of Davenport*

Nevertheless, the transformation of the differential equation into equation (1) gives a very interesting result.

Theorem [Davenport, 1984c (see also 1985a)]. *Let A be a class of functions, containing f and g, and let us suppose that the equation $y' + fy = g$ has an elementary solution over A. Then:*

> either $e^{\int f}$ *is algebraic over A, and in this case the theory of integration ought to be able to determine y;*
>
> or y *belongs to A.*

Proof. If $e^{\int f}$ is not algebraic over A, then this function is transcendental over A. We put $B = A(\int f)$. Either $e^{\int f}$ is algebraic over B, or it is transcendental over B. The only way in which $e^{\int f}$ can be algebraic over B is for $\int f$ to be a sum of logarithms with rational coefficients. But in this case, $e^{\int f}$ is algebraic over A .

---

* The reader may notice that this analysis is very similar to the analysis of the denominator. This similarity is not the result of pure chance — in fact the amount by which the degree of Y is greater than that of D is the multiplicity of $\hat{x}$ in the denominator of y, after carrying out the transformation $\hat{x} = 1/x$. Davenport [1984a, b] analyses this point.

The only other possibility to be considered is that $e^{\int f}$ is transcendental over B. We supposed y elementary over A, and *a fortiori*, y elementary over B. Now the quotient of two elementary functions is an elementary function, and thus

$$ye^{\int f} = \int g e^{\int f}$$

is elementary over B. But the decomposition lemma for integrals of exponential functions implies that this integral has the form $he^{\int f}$, where h belongs to B. Then $y \in B$, and y is the solution to a problem of Risch: in fact to Risch's problem for the original equation $y' + fy = g$. But f and g belong to A, and therefore the solution of Risch's problem belongs to the same class.

In fact, this theorem asserts that there are only two ways of calculating an elementary solution of a first order differential equation: either we find a solution to the homogeneous problem, or we find a solution in the same class as the coefficients. Moreover, it suffices to look for the homogeneous algebraic solutions over the class defined by the coefficients.

5.2.2 Second order equations

Every first order equation has, as we have seen, a solution which can be expressed in terms of integrals and of exponentials, and the interesting problem about these equations is to determine whether or not the solution is elementary. For second order equations, it is no longer obvious (or true) that every solution can be expressed in terms of integrals and of exponentials. To be able to state exact theorems, we need a precise definition of this concept.

Definition. *Let K be a field of functions. The function θ is a Liouvillian generator over K if it is:*
(a) *algebraic over K, that is if θ satisfies a polynomial equation with coefficients in K;*
(b) *θ is an exponential over K, that is if there is an η in K such that $\theta' = \eta'\theta$, which is an algebraic way of saying that $\theta = \exp \eta$;*
(c) *θ is an integral over K, that is if there is an η in K such that $\theta' = \eta$, which is an algebraic way of saying that $\theta = \int \eta$.*

Definition. *Let K be a field of functions. An over-field $K(\theta_1, \ldots, \theta_n)$ of K is called a field of Liouvillian functions over K if each θ_i is a Liouvillian generator over K. A function is Liouvillian over K if it belongs to a Liouvillian field of functions over K.*

If we omit K, we imply $\mathbf{C}(x)$: the field of the rational functions.

This definition is more general than the definition of "elementary", for every logarithm is an integral. As we have already said, this includes the trigonometrical functions. We use the notation $K(\text{liou})$ for the class of Liouvillian functions over K.

Theorem [Kovacic 1977, 1986]. *There is an algorithm which, given a second order linear differential equation, $y'' + ay' + b = 0$ with a and b rational functions in x,*

> *either finds two Liouvillian solutions such that each solution is a linear combination with constant coefficients of these solutions,*

> or *proves that there is no Liouvillian solution (except zero).*

This theorem and the resulting algorithm are quite complicated, and we cannot give all the details here. It is known that the transformation $z = e^{\int a/2} y$ reduces this equation to $z'' + (b - a^2/4 - a'/2)z = 0$. Kovacic has proved that this equation can be solved in four different ways.

(1) There is a solution of the form $e^{\int f}$, where f is a rational function. In this case, the differential operator factorises, and we get a first order equation, the solutions of which are always Liouvillian.

(2) The first case is not satisfied, but there is a solution of the form $e^{\int f}$, where f satisfies a quadratic equation with rational functions as coefficients. In this case, the differential operator factorises and we get a first order equation, the solutions of which are always Liouvillian.

(3) The first two cases are not satisfied, but there is a non-zero Liouvillian solution. In this case, every solution is an algebraic function.

(4) The non-zero solutions are not Liouvillian.

An example of the use of Kovacic's algorithm is the equation

$$y'' = \frac{4x^6 - 8x^5 + 12x^4 + 4x^3 + 7x^2 - 5x + 1}{4x^4} y,$$

for which one solution is

$$y = e^{\frac{-3}{2} \log x + \log(x^2 - 1) + \frac{1}{2}x^2 - 1 - 1/x}$$

$$= x^{-3/2}(x^2 - 1)e^{x^2/2 - x - 1/x}.$$

Kovacic's method proves also that Bessel's equation, which can be written as

$$y'' = \left(\frac{4n^2 - 1}{4x^2} - 1 \right) y,$$

only has elementary solutions when $2n$ is an odd integer, and finds solutions when n meets this requirement. This algorithm has been implemented in MACSYMA (it is a matter of regret that this implementation is not distributed with MACSYMA), and seems to work quite well [Saunders, 1981].

In the case of second degree equations, the same methods which we used for first degree equations can be used for solving equations with a second term (see Davenport [1984c] and Davenport [1985a]) but in fact the results are much more general, and will be dealt with in the next section.

5.2.3 General order equations

The situation for general order equations is, in theory, as well understood as for second order equations, even if the algorithms are not as clear.

5.2.3.1 *Homogeneous equations*

We shall first consider generalisations of Kovacic's algorithm.

Theorem [Singer, 1981]. *There is an algorithm which, given a linear differential equation of any order, the coefficients of which are rational or algebraic functions:*

    either finds a Liouvillian solution;

    or   *proves that there is none.*

If this algorithm finds a Liouvillian solution, the differential operator (which can be supposed of order n) factorises into an operator of order one, for which the solution found is a solution, and an operator of order $n - 1$, the coefficients of which are likewise rational or algebraic functions. We can apply the algorithm recursively to the latter to find all the Liouvillian solutions.

This algorithm is more general than that of Kovacic for second order equations, not only because of the generalisation of order, but also because it allows algebraic functions as coefficients. However, no-one has implemented this algorithm and it seems to be quite complicated and lengthy.

Singer [1985] has also found an algorithm like the previous one, but it looks for solutions which can be expressed in terms of solutions for second order equations and Liouvillian functions. Here again the algorithm seems quite lengthy.

5.2.3.2 *Inhomogeneous equations*

Here we look at generalisations of Davenport's theorem stated for linear first order equations. There are essentially three possibilities:

(1) There is an elementary solution;

(2) There is a Liouvillian solution, which is not elementary;

(3) There is no Liouvillian solution.

The following theorem lets us distinguish between the first case and the other two by looking for.algebraic solutions to the homogeneous equation.

Theorem 1 [Singer and Davenport, 1985, 1986]. *Let A be a class of functions containing the coefficients of a differential linear operator L, let g be an element of A, and let us suppose that the equation $L(y) = g$ has an elementary solution over A. Then:*

 either $L(y) = 0$ *has an algebraic solution over A;*
 or *y belongs to A .*

This theorem is a corollary of the following result, which lets us distinguish between the first two cases and the last one, by looking for Liouvillian solutions (of a special kind) over A.

Theorem 2 [Singer and Davenport, 1985, 1986]. *Let A be a class of functions, which contains the coefficients of a differential linear operator L, let g be an element of A, and let us suppose that the equation $L(y) = g$ has a Liouvillian solution over A. Then:*

 case 1) either $L(y) = 0$ *has a solution* $e^{\int z}$ *with z algebraic over A;*
 case 2) or y belongs to A .

Moreover, there is an algorithm for the case when A is a field of algebraic functions, which decides on the case in theorem 2, and which reduces the problem of theorem 1 to a problem in algebraic geometry. This algorithm is based on Singer's algorithm (or Kovacic's, as the case may be) for homogeneous equations. But this is quite a new subject, and there are many problems still to be solved and algorithms to be implemented and improved.

Solutions in finite form to a differential equation are very useful, if there are any. But there are no such solutions to most differential equations in physics. That is why the following section is important, for we study a method for determining solutions in series to these equations.

5.3 ASYMPTOTIC SOLUTIONS OF O.D.E.S

5.3.1 Motivation and history

The aim of this part of the book is to describe some recent developments[*] in the algorithmic methods needed for the "solution" of linear differential equations. Note that here "solution" means "solution in series".

---

 [*] This research is directed by J. Della Dora in the Computer Algebra

We shall only consider equations of the form:

$$a_n(x)(y)^{(n)} + a_{n-1}(x)(y)^{(n-1)} + \ldots + a_0(x)y = 0 \qquad (1)$$

where it is always supposed that the a_i are polynomials with complex coefficients (we shall discuss this hypothesis later), with no common factor.

Of course, differential equations such as (1) have been the subject of innumerable studies. Ever since the first papers by Gauss in 1812 and those of Kummer (1834), most great mathematicians have worked on solutions to these equations in **C**. We must mention the papers of Riemann (1857), Weierstrass (1856), Cauchy (1835–1840), before passing on to the fundamental work of Fuchs (1865), Frobenius (1873), Poincaré (1881), Birkhoff (1909), to name only the most important ones. Today these studies have been taken up again by P. Deligne (1976), B. Malgrange (1980) and J.P. Ramis (1981) from the theoretical standpoint.

Why this interest in equations such as (1) ?

There are many answers:
1) obvious theoretical interest,
2) enormous practical interest — we quote just a few applications of linear differential equations —
   solution by separation of variables of problems with partial derivatives
   solution of eigenvalue problems (Sturm-Liouville problems),
   generation of numerous special functions etc....

What can we hope to contribute to such a branch of mathematics?

Firstly, to be able for the first time to generate without error the algebraic solutions to these equations. The result is extremely important in its own right as we shall show later, for it enables us to deal with theoretical problems previously inaccessible. A second application, which we shall not consider here, is to provide a "generator" of special functions (numerical values, precise index characterisitics ...). This second set of problems alone requires an enormous amount of work, reviewing all that has already been done in this field. Finally, we have to consider that this subject is a test case — if software for solving algebraically *ordinary* differential equations

group of the Laboratory TIM3 at the IMAG Institute in Grenoble, the members of which are C. Dicrescenzo, A. Hilali, E. Tournier, A. Wazner, H. Zejli-Najid. The work is carried out in close collaboration with the University of Strasbourg (J.P. Ramis, F. Richard, J. Thoman), and with the Fourier Institute in Grenoble (D. Duval, B. Malgrange).

were found, it would open the door to work on software for solving *partial* differential equations. The future is rich in problems.

5.3.2 Classification of singularities

Let us consider an equation such as (1). The singularities of the solutions of this equation are localised at the zeros of $a_n(x) = 0$, or possibly at infinity.

By a translation such as $x \to x - \alpha$ we can reduce the problem, for zeros at a finite distance, to a singularity at the origin (the substitution $x \to 1/x$ similarly reduces the singularity at infinity to one at zero). Classically, we suppose that (1) has a singularity at zero.

This way of proceeding is in fact erroneous. For we know that, away from the singularities, the solutions are analytic functions and therefore very regular. Now, choosing a root of a polynomial leads, whatever numerical method is used (Newton, ...), to an approximation to this root and, by a numerical translation, we shall place ourselves at a regular point, and therefore we shall lose all our information about the singularity. We cannot proceed thus.

We outline briefly a way of getting out of this problem (a way which of course requires great changes of attitude by the calculator). First of all for the sake of simplicity we can suppose that the coefficients of a_n are rational numbers (therefore a_n belongs to $\mathbf{Q}[x]$) and furthermore that they are integers.

Here the roots α of a_n are not determined by numerical values (which would of course be only approximations in most cases), but simply by the computational rule "$a_n(\alpha) = 0$", which is true for all the roots of a_n. Then performing the translation $x \to x - \alpha$ has a very precise meaning, which we illustrate by an example. The new operator we obtain has its coefficients in $(\mathbf{Q}[x]/a_n)[y]$.

When we look at the differential operator

$$L(y) = (x^2 + 2)y'' + y$$

we see that its coefficients belong to $\mathbf{Z}[x]$. Let α denote a root of $x^2 + 2$; α therefore satisfies $\alpha^2 = -2$ and that is the only information we can use. Let us perform the translation

$$z = x - \alpha$$

then

$$z^2 = x^2 - 2\alpha * x + \alpha^2$$
$$= x^2 - 2\alpha * x - 2$$
$$= x^2 - 2\alpha(z + \alpha) - 2$$
$$= x^2 - 2\alpha * z + 2$$

therefore $x^2 + 2 = z^2 + 2\alpha z$ and the equation becomes

$$(z^2 + 2\alpha * z)y'' + y = L(y)$$

that is:

$$z(z + 2\alpha)y'' + y = L(y).$$

We have indeed madperformed a translation which reduces the singularity to zero. It is worth noting that there is no need to repeat this calculation for the other root.

We can therefore suppose from now on that the origin is a root of a_n, that is $a_n(0) = 0$. We have already supposed that the a_i have no common factor, therefore there is at least one index j such that $a_j(0)$ differs from 0.

To classify the singularities, we need another concept, that of
Newton's polygon of a differential operator.

We give an example to explain this.

Let the operator be

$$L = x^5 \partial^4 + x^3 \partial^3 + 2x^3 \partial^2 + \partial + 1,$$

where we write $\partial = \dfrac{d}{dx}$.

If $x^i \partial^j$ is a differential monomial belonging to L, we associate to it the point in $\mathbf{Z}^2$:

$$(j, i - j).$$

Therefore we ought to consider the points in $\mathbf{Z}^2$:

$$(4, 1); (3, 0); (2, 1); (1, -1); (0, 0).$$

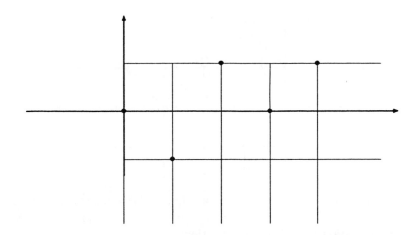

S will denote the set of points of $\mathbf{Z}^2$ associated with the monomials of L. Then, for a point (a, b) thus found, we consider the quadrant :

$$Qt(a, b) = \{(x, y) \in \mathbf{R}^2; x <= a; y >= b\}.$$

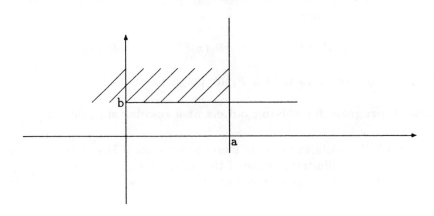

We define

$$QT = \bigcup_{(a,b) \in S} Qt(a, b).$$

Newton's polygon is the frontier of the convex hull of QT.

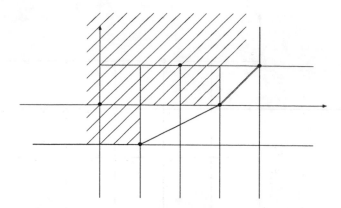

Then we can state Fuchs' characterisation:

We shall say that 0 is a regular singularity for L if the Newton polygon of L has a single slope and if this slope is zero. Otherwise we shall say that the singularity is irregular.

Note that in the classic development [Ince, 1956], other criteria are used to define the nature of the singularities. Our definition has the advantage of greater simplicity. It is easy to see that, if 0 is a regular singularity, we can write L in canonical form

$$L = x^n R_0(x)\partial^n + x^{n-1} R_1(x)\partial^{n-1} + \dots + R_n(x)$$

where $R_0(0)$ is non-zero and the R_i are polynomials.

5.3.3 A program for solving o.d.e.s at a regular singularity

We shall explain this case in more detail, since it is simpler than the general case and illustrates some of the difficulties encountered. The approach outlined here follows a classic algorithm of Frobenius (other algorithms are possible).

(1) We look at the action of L on a symbolic power x^λ, and see that

$$L(x^\lambda) = x^\lambda f(x, \lambda)$$

where

$$f(x, \lambda) = \sum_{j=0}^{m} f_j(\lambda)x^j \tag{1}$$

with f_j a polynomial in the variable λ, and m a function of the degrees of $R_0, \ldots, R_n$.

(2) We can look formally for solutions of the form

$$x^\lambda \sum_{j=0}^{+\infty} g_j x^j \quad \text{with } g_0 \neq 0.$$

The linearity of L lets us write

$$L(x^\lambda \sum_{j=0}^{+\infty} g_j x^j) = \sum_{j=0}^{+\infty} g_j L(x^{\lambda+j}).$$

(3) Using (1) it is easily seen that

$$L(x^\lambda \sum_{j=0}^{+\infty} g_j x^j) = x^\lambda \sum_{j=0}^{+\infty} g_j \left(\sum_{i=0}^{k} f_i(\lambda + j)x^i\right),$$

and therefore $x^\lambda \sum g_j x^j$ will be a solution of L if and only if the following infinite system is satisfied:

$$\begin{pmatrix} f_0(\lambda) & 0 & \cdots & \cdots & \cdots \\ f_1(\lambda) & f_0(\lambda+1) & 0 & \cdots & \cdots \\ \vdots & \vdots & \ddots & \vdots & \ddots \\ f_k(\lambda) & f_{k-1}(\lambda+1) & \cdots & f_0(\lambda+k) & \ddots \\ 0 & \ddots & \ddots & \ddots & \ddots \end{pmatrix} \begin{pmatrix} g_0 \\ g_1 \\ \vdots \\ g_k \\ \vdots \end{pmatrix} = 0.$$

This infinite system can be broken down into two parts:
— the initial conditions part

$$f_0(\lambda)g_0 = 0$$
$$f_1(\lambda)g_0 + f_0(\lambda + 1)g_1 = 0$$
$$f_2(\lambda)g_0 + f_1(\lambda + 1)g_1 + f_0(\lambda + 2)g_2 = 0$$
$$\ldots\ldots\ldots\ldots\ldots$$
$$\ldots\ldots\ldots\ldots\ldots$$
$$\ldots\ldots\ldots\ldots\ldots$$
$$f_{(k-1)}(\lambda)g_0 + \cdots\cdots + f_0(\lambda + k - 1)g_{k-1} = 0$$

— the other part is in fact a linear recurrence equation

$$f_k(\lambda + n)g_n + \ldots\ldots\ldots + f_0(\lambda + k + n)g_{n+k} = 0 \quad \text{for } n \geq 0.$$

We see that the g_i can be expressed algebraically as rational fractions in λ, say for example:

$$g_1 = \frac{-f_1(\lambda)}{f_0(\lambda + 1)g_0}.$$

It is important to note here that the g_i will be obtained by exact calculations.

The only difficulty is the first equation:

$$f_0(\lambda)g_0 = 0.$$

Since g_0 cannot be set equal to zero, the only solution is to choose λ such that $f_0(\lambda) = 0$, but here we come up against a problem we have already encountered, made worse by the difficulty which arises from the fact that f_0 is an element of $(\mathbf{Q}[x]/(a_n))[y]$, and we have therefore to make a further extension, and work in

$$(\mathbf{Q}[x]/(a_n)[y]).$$

We shall not go any deeper into this delicate question, but shall conclude with the remark that certain denominators of the g_i can be zero. In its current state, the Frobenius program can deal with the second difficulty mentioned, thanks to the recent work of Dicrescenzo and Duval [1985].

5.3.4 The general structure of the program

The program DESIR is made up of the following basic modules:

FROBENIUS, which we have just outlined, lets us deal with the case of regular singularities.

NEWTON, which lets us deal with the case of irregular singularities. We state here the basic idea of the method, starting from the following simple example: let the differential operator of order $n = 2$ be

$$L = x^2\delta + \theta(x) \quad , \qquad \theta \in k[x], \quad \theta(0) \neq 0$$

We construct the Newton polygon associated with L:

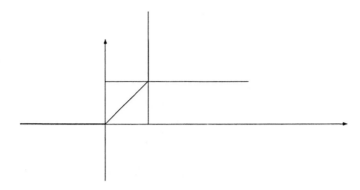

We see that in this case the Newton polygon has a slope $\lambda = 1$. It is well known that we can find n linearly independent algebraic solutions: $u_1, u_2, \dots u_n$ of $L(u) = 0$ which are written

$$u_i = e^{Q_i(x)} v(x),$$

where $Q_i(x)$ is a polynomial in $x^{(-1/q_i)}$, with $q_i \in \mathbf{Z}$. In our example, the solutions will therefore be of the form:

$$u = e^{(a/x^\lambda)} v(x) \quad \text{with } \lambda = 1.$$

With the operator L and with the slope $\lambda = 1$ we associate a new operator L_a obtained by the change of variable

$$u = e^{\frac{a}{x}} v(x),$$

which gives

$$L(e^{\frac{a}{x}} v(x)) = x^2 \left(-\frac{a}{x^2} e^{\frac{a}{x}} v(x) + e^{\frac{a}{x}} \delta v(x) \right) + \theta(x) e^{\frac{a}{x}} v(x)$$
$$= e^{\frac{a}{x}} (x^2 \delta + (\theta(x) - a)) v(x),$$

therefore we make the operator $L_a = x^2 \delta + (\theta(x) - a)$ correspond to L.

The term of L_a corresponding to the point s_0 of the polygon is the polynomial $P(a) = \theta(0) - a$. We call the equation $P(a) = 0$ the characteristic equation of L_a. Therefore if we give a the value of the root of the characteristic equation, say:

$$a = \theta(0)$$

the point s_0 disappears, and the Newton polygon becomes:

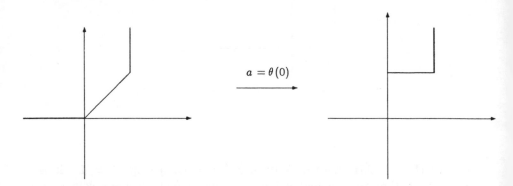

That is a Newton polygon which characterises a regular singularity whose solutions we know how to determine by the Frobenius method.

Therefore the general solution will be of the form:

$$e^{(a^*/x^\lambda)}\Psi(x),$$

where $\Psi(x)$ is a solution given by the Frobenius algorithm, and a^* is a root of the characteristic equation.

The example given here is that of the simple generic case, that is the case where a^* is a simple root of the characteristic equation, and where we get a zero slope after just one change of variable.

A complete account would have to consider the nature of the roots (simple, multiple, differing from one another by an integer) of the characteristic equation. For a detailed study of these various cases, we refer the reader to Della Dora *et al.* [1982], and to paragraph 5.3.5 of this chapter where we give several examples of DESIR.

The other modules

Associated with the two FROBENIUS and NEWTON modules which deal with differential equations, is a module which treats the case of differential systems [Hilali, 1983, 1987]:

$$\dot{X} = AX$$

where A is meromorphic at the origin.

There are also four fundamental modules being developed, viz:

1) an arithmetical module which makes it possible to manipulate the various algebraic extensions required [Dicrescenzo and Duval, 1984, 1985], [Duval, 1987b];

2) a module for numerical resummation, which makes it possible to calculate special functions [Ramis and Thomann, 1980];

3) a module for graphical visualisation [Richard, 1986];

4) a module for treating linear difference equations [Della Dora and Tournier, 1984, 1986], [Tournier, 1987].

Some of this work is already quite advanced, and for more details we efer to the papers cited.

.3.5 Some examples treated by "DESIR"

)ESIR is a program developed under the REDUCE system. The examples iven below have been implemented on the Bull DPS8 computer with the Iultics operating system. In these examples there are:

- the call which is made by the command : `codedesir()`;

- the interactive dialogue for introducing data;

- printing solutions of the differential equation with the number of terms required;

- verification of the correctness of this solution by putting it back into the given equation, the answer is then an expression, all of whose terms are of degree higher than that of the solution.

.3.5.1 *Examples with Bessel's equation*

```
›dedesir();
```

```
            *****  INTRODUCTION OF THE DATA  *****
```

The equation has the form

```
    a(1,0)(x)d^0 + a(1,1)(x)d^1 + .... + a(1,n)(x)d^n = 0
```

rder of the equation ?

Give the coefficients   a(1,j)(x)
*2-nu**2; x; x**2;

$$a(1,0) = - nu^2 + x^2$$

$$a(1,1) = x$$

$$a(1,2) = x^2$$

correction ? (yes; / no;)
no;

number of terms wanted for the solution ?
10;

***** READING FINISHED *****

***** EXECUTION *****

THE 2 SOLUTIONS CALCULATED ARE IN THE TABLE

soldif(1,j) with j = 1,2

==============

SOLUTION No 1

==============

(x^nu *(122880*nu^5 - 30720*nu^4*x^2 + 1843200*nu^4 + 3840*nu^3*x^4 - 430080*nu^3*x^2

+ 10444800*nu^3 - 320*nu^2*x^6 + 46080*nu^2*x^4 - 2181120*nu^2*x^2 +

27648000*nu^2 + 20*nu*x^8 - 2880*nu*x^6 + 180480*nu*x^4 - 4730880*nu*x^2

+ 33669120*nu - x^10 + 100*x^8 - 6400*x^6 + 230400*x^4 - 3686400*x^2 +

14745600))/(122880*(nu^5 + 15*nu^4 + 85*nu^3 + 225*nu^2 + 274*nu + 120))

==============

SOLUTION No 2

==============

(122880*nu^5 + 30720*nu^4*x^2 - 1843200*nu^4 + 3840*nu^3*x^4 - 430080*nu^3*x^2 +

$$10444800*nu^3 + 320*nu^2*x^6 - 46080*nu^2*x^4 + 2181120*nu^2*x^2 - 27648000*nu^2$$

$$+ 20*nu*x^8 - 2880*nu*x^6 + 180480*nu*x^4 - 4730880*nu*x^2 + 33669120*nu +$$

$$x^{10} - 100*x^8 + 6400*x^6 - 230400*x^4 + 3686400*x^2 - 14745600)/(122880*x^{nu} *($$

$$nu^5 - 15*nu^4 + 85*nu^3 - 225*nu^2 + 274*nu - 120))$$

We can check the validity of each solution by calling `solvalide(j)` which puts back the Jth solution into the equation (or equations when a system is being considered).

`solvalide(1);`

EQUATION 1

$$(- x^{nu} *x^{12})/(122880*(nu^5 + 15*nu^4 + 85*nu^3 + 225*nu^2 + 274*nu + 120))$$

`solvalide(2);`

EQUATION 1

$$x^{12} /(122880*x^{nu} *(nu^5 - 15*nu^4 + 85*nu^3 - 225*nu^2 + 274*nu - 120))$$

`sub(nu=0, soldif(1,1));`   % 1st solution for nu=0 ;

$$(- x^{10} + 100*x^8 - 6400*x^6 + 230400*x^4 - 3686400*x^2 + 14745600)/14745600$$

`sub(nu=0, soldif(1,2);`   % 2nd solution for nu=0 ;

$$(- x^{10} + 100*x^8 - 6400*x^6 + 230400*x^4 - 3686400*x^2 + 14745600)/14745600$$

`sub(nu=1,soldif(1,1));`   % 1st solution for nu=1 ;

$$(x*(- x^{10} + 120*x^8 - 9600*x^6 + 460800*x^4 - 11059200*x^2 + 88473600))/$$

$$88473600$$

```
sub(nu=1,soldif(1,2));          % 2nd solution for nu=1 :
                                   this 2nd solution cannot be calculated
                                   for  nu > 0 ;
***** zero denominator

errornil

sub(nu=-1,soldif(1,2))n;        % but it can be calculated for nu<0;

        10        8          6            4              2
(x*(- x    + 120*x  - 9600*x   + 460800*x  - 11059200*x  + 88473600))/

88473600

sub(nu=2,soldif(1,1));          % 1st solution for nu=2 ;

   2     10        8           6           4             2
(x *(- x    + 140*x  - 13440*x   + 806400*x  - 25804800*x  + 309657600))/

309657600



codedesir();
```

```
                     *****  INTRODUCTION OF THE DATA  *****


  The equation has the form

        a(1,0)(x)d^0 + a(1,1)(x)d^1 + .... + a(1,n)(x)d^n = 0

 order of the equation ?
2;
  Give the coefficients    a(1,j)(x)
x**2;x;x**2 ;
            2
a(1,0) = x

a(1,1) = x

            2
a(1,2) = x

correction ? ( yes; / no; )
no;

number of terms wanted for the solution ?
10;


                  *****  READING FINISHED  *****
```

*****   EXECUTION   *****

THE 2 SOLUTIONS CALCULATED ARE IN THE TABLE

soldif(1,j)   with j = 1,...2

==============

SOLUTION No 1

==============

$$(-x^{10} + 100*x^8 - 6400*x^6 + 230400*x^4 - 3686400*x^2 + 14745600)/14745600$$

==============

SOLUTION No 2

==============

$$(-60*\log(x)*x^{10} + 6000*\log(x)*x^8 - 384000*\log(x)*x^6 + 13824000*\log(x)*x^4$$
$$- 221184000*\log(x)*x^2 + 884736000*\log(x) + 137*x^{10} - 12500*x^8 + 704000*x^6$$
$$- 20736000*x^4 + 221184000*x^2)/884736000$$

We can check the validity of each solution by calling `solvalide(j)` which puts the Jth solution back into the equation.

`solvalide(1);`

EQUATION 1

$$(- x^{12})/14745600$$

`solvalide(2);`

EQUATION 1

$$(x^{12} *(- 60*\log(x) + 137))/884736000$$

5.3.5.2 *Another example*

`codedesir();`

```
                 *****  INTRODUCTION OF THE DATA  *****

  Do you want to treat a system? (yes / no)
no;
  The equation has the form

        a(1,0)(x)d^0 + a(1,1)(x)d^1 + .... + a(1,n)(x)d^n = 0

 order of the equation ?
4;
  Give the coefficients    a(1,j)(x)
x+1 ;
2*x**2*(x+1);
x**4;
5*x**7/2;
x**10;
a(1,0) = x + 1

             2
a(1,1) = 2*x *(x + 1)

             4
a(1,2) = x

             7
a(1,3) = (5*x )/2

             10
a(1,4) = x

correction ? ( yes; / no; )
no;

number of terms wanted for the solution ?
4;


                 *****  READING FINISHED  *****



                 *****     EXECUTION     *****

  THE 4 SOLUTIONS CALCULATED ARE IN THE TABLE

                 soldif(1,j)   with j = 1,...4

  ==============

  SOLUTION No 1

  ==============

  ((sqrt(x)+sqrt(6)*x)/(sqrt(x)*x))
 (e                                        *(- 63868560*sqrt(x)*x - 398592*sqrt(
```

$$x) + 174069763*sqrt(6)*x^2 + 2641824*sqrt(6)*x + 4608*sqrt(6)))/(4608*$$

$$sqrt(6)*x^4)$$

==============

SOLUTION No 2

==============

$$(e^{(1/x)}*(63868560*sqrt(x)*x + 398592*sqrt(x) + 174069763*sqrt(6)*x^2 +$$

$$2641824*sqrt(6)*x + 4608*sqrt(6)))/(4608*e^{(sqrt(6)/sqrt(x))}*sqrt(6$$

$$)*x^4)$$

==============

SOLUTION No 3

==============

$$(x^{(1/54)}*e^{((2*x + 3)/(3*x^2))}*x^{10}*(863316799848061*x^4 - 188871635401200*x^3$$

$$+ 41488989484320*x^2 - 7344803723904*x + 1338925209984))/1338925209984$$

==============

SOLUTION No 4

==============

$$(x^{(13/27)}*e^{(1/(4*x^2))}*x^{14}*(90412648939865456*x^4 + 2632462344749808*x^3 +$$

$$62680827168288*x^2 + 1090631723256*x + 10460353203))/(10460353203*$$

$$e^{(8/(3*x))})$$

As we have seen in the above examples, we can check the validity of each solution by calling the procedure solvalide(j) which puts the Jth solution back into the equation.

```
solvalide(1);
EQUATION 1
```

$$
(e^{(sqrt(x)^{(-1)}*sqrt(6) + x^{(-1)})} *(167106972480*sqrt(x)*sqrt(6)*x^4 +
$$

$$
464361708855*sqrt(x)*sqrt(6)*x^3 + 137967753774*sqrt(x)*sqrt(6)*x^2 -
$$

$$
128924853089*sqrt(x)*sqrt(6)*x - 40445876310*sqrt(x)*sqrt(6) +
$$

$$
696161071305*x^4 + 751890166596*x^3 - 110777578191*x^2 - 170415446928*x))/
$$

$$
(36864*sqrt(x)*sqrt(6))
$$

Appendix. Algebraic background

A.1 SQUARE-FREE DECOMPOSITION

In this section we consider a polynomial p of $R[x]$, where R is an integral ring of zero characteristic* (for example, the ring $\mathbf{Z}$ of integers). It is possible that p has multiple factors, that is that there is a polynomial q such that q^2 divides p (perhaps that q^3 or a higher power divides p, but in this case it is still true that q^2 divides p).

Obviously, we can find all the multiple factors by doing a complete factorisation of p, but there is a very much simpler way, which we now describe. It suffices to consider monic p, that is with 1 as leading coefficient. Suppose that p is factorised into a product of linear factors:

$$p = \prod_{i=1}^{n} (x - a_i)^{n_i},$$

where the a_i can be algebraic quantities over R (but we only do this factorisation for the proof).

The derivative of p, which is calculated purely algebraically starting from the coefficients of p, is then

$$p' = \sum_{i=1}^{n} \left(n_i (x - a_i)^{n_i - 1} \prod_{\substack{j=1 \\ i \neq j}}^{n} (x - a_j)^{n_j} \right),$$

---

* We make this hypothesis in order to exclude the case of $x^p + 1$ over the field with p elements: this polynomial is irreducible but its derivative is zero. This and similar cases are not very difficult to handle — see, for example, Appendix 3 of Davenport [1981].

It is obvious that, for every i, $(x - a_i)^{n_i-1}$ divides p and p'. Moreover, every polynomial which divides p is a product of the $(x - a_i)$, to a power less than or equal to n_i. Therefore the g.c.d. of p and p' is almost determined: it is the product of the $(x - a_i)$, to a power which is $n_i - 1$ or n_i. But this power cannot be n_i, for $(x - a_i)^{n_i}$ divides all the terms of p' except one, and cannot therefore divide p'. So we have proved that

$$\gcd(p, p') = \prod_{i=1}^{n} (x - a_i)^{n_i-1},$$

Thus, $p/\gcd(p, p') = \prod_{i=1}^{n} (x - a_i)$. For the sake of brevity let us call this object q. Then

$$\gcd(q, \gcd(p, p')) = \prod_{\substack{i=1 \\ n_i > 1}}^{n} (x - a_i),$$

from which we deduce

$$\frac{q}{\gcd(q, \gcd(p, p'))} = \prod_{\substack{i=1 \\ n_i = 1}}^{n} (x - a_i).$$

The lefthand side of this equation is the product of all the non-multiple factors of p, and we have shown how to calculate it using only the operations of derivation, (exact) division and g.c.d. These operations do not make us leave $R[x]$, which implies that this product can be calculated in $R[x]$, and by a fairly efficient method.

In addition, we have calculated as an intermediary result $\gcd(p, p')$, which has the same factors as the multiple factors of p, but with their powers reduced by one. If the same calculation we have just applied to p is applied to $\gcd(p, p')$, we are calculating all the factors of $\gcd(p, p')$ of multiplicity one, that is the factors of p of multiplicity two. And similarly for the factors of multiplicity three, ...

That is, by quite simple calculations in $R[x]$, we can break down p in the form $\prod p_i^i$, where p_i is the product of all the factors of p of multiplicity i. In this decomposition of p, each p_i is itself without multiple factors, and the p_i are relatively prime. This decomposition is called the *square-free decomposition* of p (also called "quadrat-frei").

The method we have shown may seem so effective that no improvement is possible but Yun [1976, 1977] has found a cunning variation, which shows that, asymptotically, the cost of calculating a square-free decomposition of p is of the same order of magnitude as the cost of calculating the

$\gcd(p, p')$. We see that all these calculations are based on the possibility of our being able to calculate the derivative of p. Calculating a square-free decomposition of an integer is very much more complicated — it is almost as expensive as factorising it.

A.2 THE EXTENDED EUCLIDEAN ALGORITHM

According to Euclid (see the historical notes in Knuth [1973]), we can calculate the g.c.d. of two integers q and r by the following method:

$$\textbf{if } |q| < |r| \textbf{ then } t := q;$$
$$q := r;$$
$$r := t;$$
$$\textbf{while } r \neq 0$$
$$\textbf{do } t := \text{remainder}(q, r);$$
$$q := r;$$
$$r := t;$$
$$\textbf{return } q;$$

(where we have used indentation to indicate the structure of the program, rather than **begin** ... **end**, and the function "remainder" calculates the remainder from the division q/r). In fact, the same algorithm holds for the polynomials in one variable over a field — it suffices to replace the meaning "absolute value" of the signs $||$ by the meaning "degree".

This algorithm can do more than the simple calculation of the g.c.d. At each stage, the values of the symbols q and r are linear combinations of the initial values. Thus, if we follow these linear combinations at each stage, we can determine not only the g.c.d., but also its representation as a linear combination of the data. Therefore we can rewrite Euclid's algorithm as follows, where the brackets [...] are used to indicate pairs of numbers and Q and R are the pairs which give the representation of the present values of q and r in terms of the initial values.

The assignment to T may seem obscure, but the definition of remainder gives t the value $q - \lfloor q/r \rfloor r$, where $\lfloor q/r \rfloor$ is the quotient of q over r.

The first value this algorithm returns is the g.c.d. of q and r (say p), and the second is a pair $[a, b]$ such that $p = aq + br$. This equality is called *Bézout's identity*, and the algorithm we have just explained which calculated it is called the *extended Euclidean algorithm*.

We said that Euclid's algorithm applies equally to polynomials in one variable. The same is true of the extended Euclidean algorithm. But there is a little snag here. Let us suppose that q and r are two polynomials with integer coefficients. Euclid's algorithm calculates their g.c.d., which is also a polynomial with integer coefficients, but it is possible that the intermediary

Extended Euclidean Algorithm

$$
\begin{aligned}
\textbf{if } |q| < |r| \textbf{ then } & t := q; \\
& q := r; \\
& r := t; \\
& Q := [0,1]; \\
& R := [1,0]; \\
\textbf{else } & Q := [1,0]; \\
& R := [0,1]; \\
\textbf{while } r \neq 0 & \\
\textbf{do } & t := \mathrm{remainder}(q,r); \\
& T := Q - \lfloor q/r \rfloor R; \\
& q := r; \\
& r := t; \\
& Q := R; \\
& R := T; \\
\textbf{return } & q \text{ and } Q;
\end{aligned}
$$

---

values are polynomials with rational coefficients. For example, if $q = x^2 - 1$ and $r = 2x^2 + 4x + 2$, then the quotient $\lfloor q/r \rfloor$ is $1/2$. After this division, our variables have the following values:

$$
q : 2x^2 + 4x + 2 \qquad r : -2x - 2
$$
$$
Q : [0,1] \qquad\qquad R : [1, -1/2].
$$

The remainder from the second division is zero, and we get the g.c.d. $-2x - 2$, which can be expressed as

$$
1(x^2 - 1) - \frac{1}{2}(2x^2 + 4x + 2).
$$

Perhaps it is more natural to make the g.c.d. monic, which then gives us as g.c.d. $x + 1$, which is expressible as

$$
\frac{-1}{2}(x^2 - 1) + \frac{1}{4}(2x^2 + 4x + 2).
$$

A.3 PARTIAL FRACTIONS

Two fractions (of numbers or of polynomials in one variable) can be added:

$$
\frac{a}{p} + \frac{b}{q} = \frac{aq/\gcd(p,q) + bp/\gcd(p,q)}{\mathrm{lcm}(p,q)}.
$$

Is it possible to transform a fraction in the other direction, that is, can we rewrite c/pq in the form $(a/p) + (b/q)$? Such a decomposition of c/pq is called a decomposition into *partial fractions*.

This is only possible if p and q are relatively prime, for otherwise the denominator of $(a/p) + (b/q)$ would be $\mathrm{lcm}(p, q)$, which differs from pq. If p and q are relatively prime, the g.c.d. is 1. By the theorem of the preceding paragraph, there are two (computable!) integers (or polynomials, depending on the case) P and Q with $Pp + Qq = 1$. Then

$$\frac{c}{pq} = \frac{c(Pp + Qq)}{pq} = \frac{cQq}{pq} + \frac{cPp}{pq} = \frac{cQ}{p} + \frac{cP}{q},$$

and we have arrived at the desired form.

In practice, c/pq is often a *proper* fraction, that is the absolute value (if it is a question of numbers, the degree if it is a question of polynomials) of c is less than that of pq. In that case, we want the partial fractions to be proper, which is not generally guaranteed by our method. But it suffices to replace cQ by the remainder after division of it by p, and cP by the remainder after its division by q.

This procedure obviously extends to the case of a more complicated denominator, always with the proviso that all the factors are relatively prime. It suffices to take out the factors successively, for example,

$$\begin{aligned}
\frac{c}{p_1 p_2 \cdots p_n} &= \frac{a_1}{p_1} + \frac{b_1}{p_2 \cdots p_n} \\
&= \frac{a_1}{p_1} + \frac{a_2}{p_2} + \frac{b_2}{p_3 \cdots p_n} \\
&= \cdots \\
&= \frac{a_1}{p_1} + \frac{a_2}{p_2} + \cdots + \frac{a_n}{p_n}.
\end{aligned}$$

In fact, there are more efficient methods than this, which separate the p_i in a more balanced fashion [Abdali *et al.*, 1977].

A.4 THE RESULTANT

It quite often happens that we have to consider whether two polynomials, which are usually relatively prime, can have a common factor in certain special cases. The basic algebraic tool for solving this problem is called the resultant. In this section we shall define this object and we shall give some properties of it. We take the case of two polynomials f and g in one variable x and with coefficients in a ring R.

We write $f = \sum_{i=0}^{n} a_i x^i$ and $g = \sum_{i=0}^{m} b_i x^i$.

Definition. *The Sylvester matrix of* f *and* g *is the matrix*

$$
\begin{pmatrix}
a_n & a_{n-1} & \cdots & a_1 & a_0 & 0 & 0 & \cdots & 0 \\
0 & a_n & a_{n-1} & \cdots & a_1 & a_0 & 0 & \cdots & 0 \\
\vdots & \ddots & \ddots & \ddots & \cdots & \ddots & \ddots & \ddots & \vdots \\
0 & \cdots & 0 & a_n & a_{n-1} & \cdots & a_1 & a_0 & 0 \\
0 & \cdots & 0 & 0 & a_n & a_{n-1} & \cdots & a_1 & a_0 \\
b_m & b_{m-1} & \cdots & b_1 & b_0 & 0 & 0 & \cdots & 0 \\
0 & b_m & b_{m-1} & \cdots & b_1 & b_0 & 0 & \cdots & 0 \\
\vdots & \ddots & \ddots & \ddots & \cdots & \ddots & \ddots & \ddots & \vdots \\
0 & \cdots & 0 & b_m & b_{m-1} & \cdots & b_1 & b_0 & 0 \\
0 & \cdots & 0 & 0 & b_m & b_{m-1} & \cdots & b_1 & b_0
\end{pmatrix}
$$

where there are m *lines constructed with the* a_i, n *lines constructed with the* b_i.

Definition. *The resultant of* f *and* g, *written* $\mathrm{Res}(f, g)$, *or* $\mathrm{Res}_x(f, g)$ *if there has to be a variable, is the determinant of this matrix.*

Well-known properties of determinants imply that the resultant belongs to R, and that $\mathrm{Res}(f, g)$ and $\mathrm{Res}(g, f)$ are equal, to within a sign. We must note that, although the resultant is defined by a determinant, this is not the best way of calculating it. Because of the special structure of the Sylvester matrix, we can consider Euclid's algorithm as Gaussian elimination in this matrix (hence the connection betwen the resultant and the g.c.d.). One can also consider the sub-resultant method as an application of the Sylvester identity (described in the section "Bareiss's algorithm") to this elimination. It is not very difficult to adapt advanced methods (such as the method of sub-resultants described in the section "Representations of polynomials", or the modular methods described in the chapter " Advanced algorithms") to the calculation of the resultant. Collins [1971] and Loos [1982] discuss this problem. We now give a version of Euclid's algorithm for calculating the resultant. We denote by $lc(p)$ the leading coefficient of the polynomial $p(x)$, by $degree(p)$ its degree, and by $remainder(p, q)$ the remainder from the division of $p(x)$ by $q(x)$. We give the algorithm in a recursive form.

Algorithm *resultant*;
Input f, g;
Output r;
$n := degree(f)$;
$m := degree(g)$;
if $n > m$ **then** $r := (-1)^{nm} resultant(g, f)$
          else $a_n := lc(f)$;
               if $n = 0$ **then** $r := a_n^m$
                    else $h := remainder(g, f)$;
                         if $h = 0$ **then** $r := 0$
                              else $p := degree(h)$;
                                   $r := a_n^{m-p} resultant(f, h)$;
     return r;

We write $h = \sum_{i=0}^{p} c_i x^i$ and $c_i = 0$ for $i > p$. This algorithm does indeed give the resultant of f and g for, when $n \le m$ and $n \neq 0$, the polynomials $x^i g - x^i h$ (for $0 \le i < n$) are linear combinations of the $x^j f$ (for $0 \le j < m$), and therefore we are not changing the determinant of the Sylvester matrix of f and g by replacing b_i by c_i for $0 \le i < m$. Now this new matrix has the form $\begin{pmatrix} A & * \\ 0 & B \end{pmatrix}$ where A is a triangular matrix with determinant a_n^{m-p} and B is the Sylvester matrix of f and h. From this algorithm we immediately get

Proposition 1. $\mathrm{Res}(f, g) = 0$ if and only if f and g have a factor in common.

It is now easy to prove the following propositions:

Proposition 2. If the α_i are the roots of f, then

$$\mathrm{Res}(f, g) = a_n^m \prod_{i=1}^{n} g(\alpha_i).$$

Proposition 3. If the β_i are the roots of g, then

$$\mathrm{Res}(f, g) = (-1)^{mn} b_m^n \prod_{i=1}^{m} f(\beta_i).$$

Proposition 4. $\mathrm{Res}(f, g) = a_n^m b_m^n \prod_{i=1}^{n} \prod_{j=1}^{m} (\alpha_i - \beta_j)$.

Proof. [Duval, 1987a]. We write the right hand sides of the three propositions as

$$R_2(f, g) = a_n^m \prod_{i=1}^n g(\alpha_i),$$

$$R_3(f, g) = (-1)^{mn} b_m^n \prod_{i=1}^m f(\beta_i),$$

$$R_4(f, g) = a_n^m b_m^n \prod_{i=1}^n \prod_{j=1}^m (\alpha_i - \beta_j).$$

It is clear that $R_2(f, g)$ and $R_3(f, g)$ are equal to $R_4(f, g)$. The three propositions are proved simultaneously, by induction on the integer $Min(n, m)$. If f and g are swapped, their resultant is multiplied by $(-1)^{nm}$, and gives $R_4(f, g)$. We can therefore suppose that $n \le m$. Moreover $R_2(f, g)$ is equal to a_n^m when $n = 0$, as is the resultant of f and g, and $R_4(f, g)$ is zero when $n > 0$ and $h = 0$, as is the resultant. It only remains to consider the case when $m \ge n > 0$ and $h \ne 0$. But then $R_2(f, g) = a_n^{m-p} R_2(f, h)$ for $g(\alpha_i) = h(\alpha_i)$ for each root α_i of f, and the algorithm shows that $\mathrm{Res}(f, g) = a_n^{m-p} \mathrm{Res}(f, h)$, from which we get the desired result.

Definition. *The* discriminant *of* f, $\mathrm{Disc}(f)$ *or* $\mathrm{Disc}_x(f)$, *is*

$$a_n^{2n-2} \prod_{i=1}^n \prod_{\substack{j=1 \\ j \ne i}}^n (\alpha_i - \alpha_j).$$

Proposition 5. $\mathrm{Res}(f, f') = a_n \mathrm{Disc}(f)$. *Moreover* $\mathrm{Disc}(f) \in R$.

Whichever way they are calculated, the resultants are often quite large. For example, if the a_i and b_i are integers, bounded by A and B respectively, the resultant is less than $(n + 1)^{m/2} (m + 1)^{n/2} A^m B^n$, but it is very often of this order of magnitude. Similarly, if the a_i and b_i are polynomials of degree α and β respectively, the degree of the resultant is bounded by $m\alpha + n\beta$. An example of this swell is the use of resultants to calculate primitive elements — see the section "Representations·of algebraic functions".

A.5 CHINESE REMAINDER THEOREM

In this section we shall treat algorithmically this well-known theorem. There are two cases in which we shall use this theorem: that of integers and that of polynomials. Of course, these cases can be regrouped in a more abstract setting (Euclidean domains), but we leave this generality to the pure mathematicians. We shall deal first of all with the case of integers, and then (more briefly) with that of polynomials.

A.5.1 Case of integers

It is simpler to deal first with the case where the two moduli are relatively prime, and then to go on to the general case.

Chinese remainder theorem (first case). *Let M and N be two relatively prime integers. For every pair (a, b) of integers, there is an integer c such that $x \equiv a \pmod{M}$ and $x \equiv b \pmod{N}$ if and only if $x \equiv c \pmod{MN}$.*

Proof. By the theory of the section "Extended Euclidean algorithm", there are two integers f and g such that $fM + gN = 1$ (here we are making the hypothesis that M and N are relatively prime). Let $c = a + (b - a)fM$.

If $x \equiv c \pmod{MN}$, then $x \equiv c \pmod{M}$. But $c \equiv a \pmod{M}$, and therefore $x \equiv a \pmod{M}$. Moreover,

$$c = a + (b - a)fM = a + (b - a)(1 - gN) \equiv a + (b - a) \pmod{N} = b,$$

and therefore $x \equiv c \pmod{MN}$ implies $x \equiv b \pmod{N}$.

In the other direction, let us suppose that $x \equiv a \pmod{M}$ and that $x \equiv b \pmod{N}$. We have shown that $c \equiv a \pmod{M}$ and $c \equiv b \pmod{N}$, therefore we must have $x \equiv c \pmod{M}$ and $\pmod{N}$. But the statement that $x \equiv c \pmod{M}$ is equivalent to asserting that "M divides $x - c$". Similarly N divides $x - c$, and therefore MN divides it. But this is equivalent to "$x \equiv c \pmod{MN}$".

Corollary. *Let $N_1, \ldots, N_n$ be integers, pairwise relatively prime. For every set $a_1, \ldots, a_n$ of integers there is an integer c such that, for every i, $x \equiv a_i \pmod{N_i}$ if and only if $x \equiv c \pmod{\prod_{i=1}^{n} N_i}$.*

The same methods can be applied to the case when M and N are not relatively prime. Although this generalisation is rarely used, we present it here, because it is often misunderstood.

(Generalised) Chinese remainder theorem. *Let M and N be two integers. For every pair (a, b) of integers:*

   either $a \not\equiv b \pmod{\gcd(M, N)}$, *in that case it is impossible to satisfy the equations $x \equiv a \pmod{M}$ and $x \equiv b \pmod{N}$;*

   or   *there is an integer c such that $x \equiv a \pmod{M}$ and $x \equiv b \pmod{N}$ if and only if $x \equiv c \pmod{\mathrm{lcm}(M, N)}$.*

Proof. The first case, $a \not\equiv b \pmod{\gcd(M, N)}$, is quite simple. If $x \equiv a \pmod{M}$, then $x \equiv a \pmod{\gcd(M, N)}$. Similarly, if $x \equiv b \pmod{N}$, then $x \equiv b \pmod{\gcd(M, N)}$. But these two equations are contradictory.

In the other case, we factorise M and N into products of prime numbers: $M = \prod_{i \in P} p_i^{m_i}$ and $N = \prod_{i \in P} p_i^{n_i}$. We now divide the set P of indices into two parts*:

$$Q = \{i : m_i < n_i\};$$
$$R = \{i : m_i \geq n_i\}.$$

Let $\hat{M} = \prod_{i \in R} p_i^{m_i}$ and $\hat{N} = \prod_{i \in Q} p_i^{n_i}$, such that

$$\hat{M}\hat{N} = \prod_{i \in P} p_i^{\max(m_i, n_i)} = \mathrm{lcm}(M, N).$$

This theorem (the case when the moduli are relatively prime) can be applied to the equations $x \equiv a \pmod{\hat{M}}$ and $x \equiv b \pmod{\hat{N}}$, in order to deduce that there is a c such that this system is equivalent to the equation $x \equiv c \pmod{\hat{M}\hat{N}}$.

We have to prove that this equation implies, not only that $x \equiv a \pmod{\hat{M}}$, but also that $x \equiv a \pmod{M}$ (and the same for N). Let us therefore suppose that $x \equiv c \pmod{\hat{M}\hat{N}}$ is satisfied. This implies that $x \equiv b \pmod{\hat{N}}$, and therefore the same equation modulo all the factors of $\hat{N}$, in particular modulo $\check{M} = \prod_{i \in Q} p_i^{m_i}$ (we recall that $m_i < n_i$ for $i \in Q$). $\check{M}$ is also a factor of M, and therefore divides $\gcd(M, N)$. Since $a \equiv b \pmod{\gcd(M, N)}$, we deduce that $a \equiv b \pmod{\check{M}}$. We already know that $x \equiv b \pmod{\check{M}}$, and therefore $x \equiv a \pmod{\check{M}}$. From this and from $x \equiv a \pmod{\hat{M}}$, we deduce $x \equiv a \pmod{\hat{M}\check{M}}$. But

$$\hat{M}\check{M} = \prod_{i \in R} p_i^{m_i} \prod_{i \in Q} p_i^{m_i} = M,$$

and therefore the equation modulo M which has to be satisfied really is satisfied. The same is true for N, and therefore the theorem is proved.

A.5.2 Case of polynomials

In this section we repeat the theory discussed in the last section, for the case of polynomials in one variable (say x) with coefficients belonging to a field K. For the sake of simplicity, we suppose that each modulus is a monic polynomial. The process of calculating the remainder of a polynomial with respect to a monic polynomial is very simple if the polynomial is of degree one (that is if it is of the form $x - A$): the remainder is simply the value of the polynomial at the point A.

---

* The distinction between $<$ and $\geq$ is not very important, provided the two sets are disjoint and contain every element of P.

Chinese remainder theorem (first case). *Let M and N be two relatively prime monic polynomials. For every pair (a, b) of polynomials there is a polynomial c such that $X \equiv a \pmod{M}$ and $X \equiv b \pmod{N}$ if and only if $X \equiv c \pmod{MN}$.*

Proof. The theory of the section "Extended Euclidean algorithm" holds also for the present case, and therefore the proof follows the same lines without any changes. There are two polynomials f and g such that $fM + gN = 1$, and we put $c = a + (b - a)fM$.

Corollary. *Let $N_1, \dots, N_n$ be polynomials relatively prime in pairs. For every set $a_1, \dots, a_n$ of polynomials, there is one polynomial c such that $X \equiv a_i \pmod{N_i}$ if and only if $X \equiv c \pmod{\prod_{i=1}^{n} N_i}$.*

Chinese remainder theorem (generalised). *Let M and N be two polynomials. For every pair (a, b) of the polynomials:*

    either   *$a \not\equiv b \pmod{\gcd(M, N)}$, in such a case it is impossible to satisfy the equations $X \equiv a \pmod{M}$ and $X \equiv b \pmod{N}$;*

    or   *there is a polynomial c such that $X \equiv a \pmod{M}$ and $X \equiv b \pmod{N}$ if and only if $X \equiv c \pmod{\operatorname{lcm}(M, N)}$.*

In the case when all the polynomials are linear, these methods are equivalent to the Lagrange interpolation, which determines the polynomial p which has the values $a_1, \dots, a_n$ at the points $N_1, \dots, N_n$, that is modulo the polynomials $(x - N_1), \dots, (x - N_n)$.

As we are using this case in modular calculation, it is useful to prove that this algorithm is especially simple. Let us suppose that a is a polynomial over K, and b a value belonging to K, and that the equations to be solved are $X \equiv a \pmod{M}$ and $X \equiv b \pmod{x - v}$ (with M and $x - v$ relatively prime, that is $M_{x=v} \neq 0$). Let V be an element of K such that the remainder from dividing VM by $x - v$ is 1, that is $V = (M_{x=v})^{-1}$. Then the solution is

$$X \equiv Z = a + MV(b - a_{x=v}) \pmod{M(x - v)}.$$

It is clear that $Z \equiv a \pmod{M}$, and the choice of V ensures that $Z \equiv b \pmod{x - v}$.

If, in addition, M is of the form $\prod (x - v_i)$, V has the special value $\prod (v - v_i)^{-1}$, and can be calculated in K.

Annex. REDUCE: a Computer Algebra system

R.1 INTRODUCTION

Over the last twenty years substantial progress has been made throughout the field of Computer Algebra, and many software systems have been produced (see also Chapter 1).

Among the major Computer Algebra systems in service today, we have chosen to give here a detailed exposition of REDUCE. Why REDUCE, when it can be argued that the most complete system today is MACSYMA? Firstly, REDUCE occupies a privileged position: it is, and has been for a long time, the most widely distributed system, and is installed in well over 1000 sites, on a wide spectrum of computers. Secondly, its PASCAL-like syntax makes it easy to learn and simple to use.

This system is the brain-child of A.C. Hearn, and the first version appeared in 1967. It has continued to develop with the far-flung collaboration of its community of "advanced" users, who have contributed many modules and facilities to the system, and will surely continue to do so.

The principal possibilities which this system offers are:
- Integer and rational "arbitrary precision" arithmetic
- Machine and "arbitrary precision" floating-point arithmetic
- Polynomial algebra in one or several variables
    - g.c.d. computations
    - factorisation of polynomials with integer coefficients
- Matrix algebra, with polynomial or symbolic coefficients
    - Determinant calculations
    - Inverse computation
    - Solution of systems of linear equations
- Calculus

- Differentiation
- Integration
- Manipulation of expressions
  - Simplification
  - Substitution

REDUCE was, from its beginning, conceived of as an interactive system. A "program" is therefore a series of free-standing commands, each of which is interpreted and evaluated by the system in the order in which the commands are entered, with the result being displayed before the next command is entered. These commands may, of course, be function calls, loops, conditional statements etc.

R.1.1 Examples of interactive use

REDUCE commands end with ";" or "$": in the second case, the result is not printed. REDUCE prompts for input by typing "1:" for the first input, and so on. The user's input follows this prompt in all the examples below*.

```
REDUCE 3.3, 28-Jun-87 ...

1: (x+y+z)**2;

 2                   2         2
X  + 2*X*Y + 2*X*Z + Y  + 2*Y*Z + Z

2: df((x+y)**3,x,2);

6*(X + Y)

3: int(e**a*x,x);

  A  2
 E *X
-------
   2

4: int(x*e**(a*x)/((a*x)**2+2*a*x+1),x);

     A*X
    E
---------------
 2
A *(A*X + 1)

5: for i:=1:40 product i;
```

---

* In future examples, we shall omit the opening and closing stages of the dialogue.

815915283247897734345611269596115894272000000000

```
6: matrix m;

7: m:=mat((a,b),(b,c))$
8: 1/m;
```

$$MAT(1,1) := \frac{C}{A*C - B^2}$$

$$MAT(1,2) := - \frac{B}{A*C - B^2}$$

$$MAT(2,1) := - \frac{B}{A*C - B^2}$$

$$MAT(2,2) := \frac{A}{A*C - B^2}$$

```
9: det(m);
```

$$A*C - B^2$$

```
10: bye;

Quitting
```

These few examples illustrate the "algebraic" mode of REDUCE. There is another mode, "symbolic", which allows one to execute arbitrary LISP instructions directly. These modes correspond to the two data types "algebraic expression" and "S-expression of LISP".

The REDUCE system is itself written in LISP, and can be seen as a hierarchy:

REDUCE	algebraic mode
RLISP	infix symbolic mode
LISP	prefix symbolic mode

It is possible to communicate between the various modes. But REDUCE is essentially designed to work in algebraic mode, and symbolic mode calculation requires a good knowledge of LISP and of the internal structure of REDUCE. Hence, in this annex, we will restrict ourselves to algebraic mode. RLISP is described by Davenport & Tournier [1983].

R.2 SYNTAX OF REDUCE

We concentrate on the syntactic elements of REDUCE in this section, and pass later to the various algebraic facilities offered.

R.2.1 Syntactic elements

A REDUCE program is, as we have already said, a sequence of simple statements evaluated in order. These statements can be declarations, commands or expressions. These expressions are composed of numbers, variables, operators, strings (delimited by "), reserved words and delimiters. While this syntactic structure is common to most high-level languages, we should note the following differences.

- **The numbers**

  The integers of REDUCE are "infinite precision", and numbers which are not exact integers are normally represented as the quotient of two integers. It is possible to use machine floating-point numbers directly, and floating-point numbers of any precision.

- **The variables**

  Each variable has a name and a type. The name must not be that of a reserved word, i.e. a keyword of the language. Most variables can take the default type, which is SCALAR. Such variables can be given any algebraic expression as value. In the absence of any such value, they have themselves as value. The reserved "variables" are:

  E – the base of the natural logarithms;

  I – the square root of -1 (but it can also be used as a loop counter);

  NIL – a synonym of zero;

  PI – the circular constant π;

  T – a synonym for "true".

- **The operators**

  These come in two types.

  - **prefix**

    A number of prefix operators with special properties are incorporated in the system: a detailed list is given in Section R.3.1. The user can also add new operators, with similar properties (see R.2.3.4).

  - **infix**

    The built-in infix operators are: := OR AND NOT NEQ = >= > < <= .   + - * / **, and the user can also add new ones (see Section R.2.3.4).

R.2.2 Expressions

Since REDUCE is a Computer Algebra system, its whole purpose is the manipulation of expressions.

R.2.2.1 *Different types of expressions*

The various kinds of expression possible in REDUCE are:

- scalar expressions (i.e. those with algebraic values), computed with +, -, *, / and ** (it is often possible to use ^ instead of **);
- equational expressions, such as A+B=C, which can be used as arguments to functions like SOLVE and SUB (the substitution function);
- integer expressions, similar to those of other languages;
- boolean expressions.

These last have the (fairly standard) syntax rules:

```
<expression> <relational operator> <expression>
<boolean function>(<arguments>)
<boolean expression> <logical operator> <boolean expression>
```

We should note that operators like AND and OR are lazy, in the sense that they do not evaluate their second argument if the first arguments suffices to determine the result (false for AND, true for OR). The boolean functions built-in are:

EVENP(U) determines if U (which must be an integer expression) is even;

FIXP(U) determines if U is an integer;

FREEOF(U,V) determines if the expression U does not contain the kernel* V;

NUMBERP(U) determines if U is a number;

ORDP(U,V) determines if the kernel U is ordered before the kernel V.

R.2.2.2 *Simplification of expressions*

As has already been mentioned (see Moses [1971] and Chapter 2), this is a major problem of Computer Algebra. Some operations (e.g. collection of common terms) are always performed. Others are controlled by various switches, which can be set by the statement** ON <switch-name>, and unset via OFF <switch-name>. There any many such switches — those which control the simplification of expressions are:

EXP causes expressions to be expanded during evaluation, and is on by default (turning it off can prevent canonical forms, as the following example shows

---

* A kernel is a simple variable, or an instance of an operator for which no substitution has been defined, e.g. sin(a). See also the discussion at the beginning of Section 7 of Chapter 2.

** Several switch names can be given in a single statement, separated by commas.

```
1: off exp;

2: a:=(x+1)**2;
```

$$A := (X + 1)^2$$

```
3: b:=x**2+2*x+1;
```

$$B := X^2 + 2*X + 1$$

```
4: if (a=b) then 1 else 0;

0
```

and is therefore not recommended for beginners);

GCD   causes the system to cancel greatest common divisors during the calcu-
      lation of rational expressions (this switch is off by default, and should
      be set on if the user wishes to be sure of having canonical forms during
      rational evaluations, as we see below:

```
1: (x**2-1)/(x-1)**2;
```

$$\frac{X^2 - 1}{X^2 - 2*X + 1}$$

```
2: on gcd;

3: (x**2-1)/(x-1)**2;
```

$$\frac{X + 1}{X - 1}$$

however, simple factors are often spotted even when GCD is not set);

LCM   causes the system to compute true least common multiples when com-
      bining fractions, even when GCD is off, thus providing most of the power
      of GCD, and is normally on;

MCD   causes the system to place the sum of rational expressions over a com-
      mon denominator, and is on by default — turning it off can prevent the
      system from having canonical, or even normal, forms, as the following
      example shows:

```
1: 1/(x-1)-1/(x+1)-2/(x**2-1);

0

2: off mcd;

3: 1/(x-1)-1/(x+1)-2/(x**2-1);

        2     -1            -1              -1
  - (2*(X  - 1)    + (X + 1)    - (X - 1)   )
```

and again changing this switch is not recommended for beginners.

FLOAT causes the system to perform calculations in floating-point according
to the machine's built-in precision. This ought not to be used by default, since it causes loss of canonical forms, incorrect computations
of greatest common divisors, etc. (see the section "respresentations of
fractions" in Chapter 2). The following example illustrates some of
these problems.

```
1: on float;

2: (3^10)*(x+1/3)^10 - (3*x+1)^10;

               9            8             7             6
  - 0.137781*(X  + 3.28571*X  + 4.71428*X  + 3.88888*X  +


              5             4             3
     2.03703*X  + 0.703702*X  + 0.160493*X  +


              2
     0.0233195*X  + 0.00195962*X + 7.25786e-05)

3: gcd((x^2-1/9)^5,(x^2+(2/3)*x+1/9)^5);

1
```

Calculation **2:** should have returned 0, and **3:** should have given
$(x + 1/3)^5$.

FLOAT causes the system to perform calculations in "arbitrary precision" floating point. The number of decimal digits to be used is set by the PRE-
CISION statement, as the following example shows:

```
1: on bigfloat;

2: precision 40;

40

3: 1/3;

0.333 33333 33333 33333 33333 33333 33333 33333 33
```

The same caveats about loss of canonical forms etc apply here as applied in the case of `on float;` above.

R.2.2.3 *List expressions*

A list* is a sequence of expressions (which may themselves be lists). Lists print as the elements themselves, surrounded by braces, as in the following examples:

```
1: l:={a,{b,c},1};
```

```
L := {A,{B,C},1}
```

```
2: second l;
```

```
{B,C}
```

Parts of a list can be selected with the operators `FIRST`, `SECOND`, `THIRD` and `REST`. A new element can be added to the front of a list with the infix operator ".". (equivalent to LISP's `CONS` function). There are also the functions `APPEND` (which takes precisely two arguments) and `REVERSE`, as below:

```
3: reverse l;
```

```
{1,{B,C},A}
```

R.2.3 Declarations.

R.2.3.1 *Plain variables*

As we have already said, a variable which has not been declared is of type `SCALAR`, and has itself as value. Variables can be explicitly declared inside blocks, to be either of type `SCALAR` or of type `INTEGER`. Both declarations give the variable a new binding, extending throughout that block and hiding any other (or global) binding, as is normal for block-structured languages. They also initialise the variable to zero. Hence the example in the box (note that the prompt appears on every line of a multi-line command) works as in most other block-structured languages.

The difference between the two is that `INTEGER` variables can only take integer values. They can therefore be handled more efficiently than general variables, since +, for example, can be directly interpreted (or compiled) as the function `PLUS` of LISP, rather than as the addition of arbitrary REDUCE expressions.

It is possible to state that variables depend on one another, for the purposes of differentiation, linear operators etc. Such dependencies are introduced via the `DEPEND` declaration, as in

---

* This feature was introduced into REDUCE with version 3.3.

Example of block structuring

```
1: begin
1:   scalar a;
1:   a:=1;
1:   begin
1:     scalar a;
1:     a:=x;
1:     write a;
1:     end;
1:   write a;
1:   end;

X

1
```

depend a,x,y;
which states that **a** depends on both the variables **x** and **y**. Dependencies can be removed by the NODEPEND declaration, so that, after
nodepend a,y;
a depends only on **x**. It should be noted that dependence is a *global* property, not subject to block-structuring.

R.2.3.2 *Asymptotic declarations*

As as mentioned in the section "Taylor series: simple method" of Chapter 2, it is often useful to be able to cause automatic truncation of terms with more than a certain *weight*, which we can think of as a kind of "smallness". There are two commands in REDUCE to do this:

- WEIGHT, whose arguments are a comma-separated list of equations <kernel>=<explicit positive integer>, which say that the kernel is to be regarded as having that particular weight (the integer has to be an explicitly quoted integral number, or a FOR loop index whose value is such a number;
- WTLEVEL, whose argument must be an explicit positive integer (if no WTLEVEL command has been issued, this defaults to 2), and which indicates the smallest integer at which monomials of that total weight are truncated (or never even computed).

There is an important implementation point which must be raised: in the current versions of REDUCE, a variable with a weight may *not* be used as a kernel (e.g. in SUB or in future WEIGHT statements. This "weightedness" property of a variable can be cleared (as will many others!) by use of the CLEAR statement.

R.2.3.3 *Array declarations*

Arrays (which are always arrays of objects of type SCALAR) are declared

with the `ARRAY` statement, as in
`array v(10),u(2,4);`
which declares a one-dimensional array with eleven elements (numbered
from 0 to 10) and a two-dimensional array with fifteen elements (the first
index ranges from 0 to 2, the second from 0 to 4). The elements of an array
are set to zero by a declaration.

Array declarations can appear anywhere in a program, but are *global*
in scope. An array can be redeclared, in which case all the previous values
are lost, and all elements are reset to zero.

It should be noted that, as arrays are global in scope, there is no way
of passing an array as a parameter to a procedure. Users who have been
troubled by this in the past should consider whether their programs are not
better viewed in terms of lists (see above).

R.2.3.4 *Operator declarations*

- **Prefix operators**
  These are declared via the `OPERATOR` declaration, as in `operator f,g;`.
  Such a declaration permits their use as "arbitrary functions", as in the
  expression `f(1)+g(a,b)`. In interactive mode, the user is prompted
  whenever an unknown function is encountered, to see whether it should
  be declared as an operator in this sense.

- **Infix operators**
  These are declared with the `INFIX` declaration, but must also be given
  a parsing precedence with the `PRECEDENCE` statement. For example,
  `1: infix cross;`

  `2: precedence cross,/;`
  declares `CROSS` to be an infix operator, whose precedence is just less
  than that of `/` (and therefore greater than that of `*`), as the following
  example shows:
  `3: a*b cross c/d;`

  ```
   B CROSS C*A
   -------------
         D
  ```

- **Linear operators**
  Operators can be declared to be linear (as functions of their first argument) with respect to the second argument, by means of the `LINEAR`
  declaration. For example, we could begin the introduction of Laplace
  transforms by stating

```
1: operator laplace;

2: linear laplace;

3: laplace(1+a/s+2/s**2,s);
```

$$2*\text{LAPLACE}(\frac{1}{s^2},S) + \text{LAPLACE}(\frac{1}{S},S)*A + \text{LAPLACE}(1,S)$$

The handling of linear operators is affected by the DEPEND declarations (*vid. sup.*) in force at the time that laplace (or any linear operator) is used.

- **Symmetry and operators**
  Operators can also be declared to be symmetric or anti-symmetric via the corresponding declarations SYMMETRIC and ANTISYMMETRIC. These cause the arguments to be arranged in REDUCE's internal order. The operator $\epsilon_{i,j,k}$ could be introduced via

```
1: operator eps;

2: antisymmetric eps;
```

  after which the following simplifications take place
```
3: eps(1,2,1);

0

4: eps(3,2,1);

EPS(3,2,1)

5: eps(1,2,3);

  - EPS(3,2,1)
```

- **Non-commutative operators**
  As has been mentioned in Chapter 2 (Section 2.8), REDUCE lets one declare that certain operators are *non-commutative*, via the NONCOM declaration. For example

```
1: operator u;

2: noncom u;

3: u(a)*u(b)-u(b)*u(a);

U(A)*U(B) - U(B)*U(A)
```

R.2.3.5 *Procedure declarations*

The syntax is

```
[<type>] PROCEDURE <name> [<parameters>]; <instruction>
```
where `<type>` is one of `ALGEBRAIC`, `SYMBOLIC` or `INTEGER`;
```
<parameters> ::= ( <name> [, <name>]* )
```
and the body is a single instruction (which may well be a block: see later). In algebraic mode the default type is `ALGEBRAIC`, and may be omitted: if the entire body contains only integer calculations, then an integer procedure is more efficient, as in

```
6: integer procedure factorial n;
6: if n=0 then 1 else n*factorial(n-1);
```

```
FACTORIAL
```

R.2.4 Commands

REDUCE does not really have the strong distinction between "statements" and "expressions" that is found in languages like PASCAL. Most "statements" are really expressions, and yield a value which can be further manipulated (as in ALGOL 68). We will describe them as statements, but shall mention also the value.

R.2.4.1 *Assignment*

```
<lhs> := <expression>;
```
where `lhs` can be a simple variable, an array element or an instance of an operator*, as in
```
1: sec(0);
```

```
SEC(0)
```

```
2: sec(0):=1;
```

```
SEC(0) := 1
```

```
3: sec(0);
```

```
1
```
The value returned is that of the `expression`, thus authorising
```
a:=b:=x+y;
```

R.2.4.2 *Instruction group*

Several instructions can be collected into one via the group construct:

---

* In particular, see the discussion of `PART` in Section R.3.2.2.

```
<< [<statement>]* <expression> >>
```
whose value is that of the last expression. Note the difference between
```
<< a:=x; b:=y >>
```
whose value is Y, and
```
<< a:=x; b:=y; >>
```
whose value is 0, that of the (null) expression at the end.

R.2.4.3 *Conditional statement*

```
IF <boolean expression> THEN <expression>
```

```
IF <boolean expression> THEN <expression> ELSE <expression>
```
The value obtained is that of the branch executed (or 0 if absent). Note that
```
if x<y then 1; else 0;
```
is illegal, since the ; has terminated the conditional command. The desired effect would have been achieved by
```
if x<y then 1 else 0;
```

R.2.4.4 *Iteration statement*

The FOR statement allows iteration over number ranges or lists:
```
FOR <var> := <start> STEP <incr> UNTIL <end> <action> <expr>
```
for numeric ranges, and
```
FOR EACH <var> IN <list> <action> <expr>
```
for lists. The construction STEP 1 UNTIL can be replaced by ":".

The various forms of action allowed are: DO, PRODUCT, SUM, COLLECT and JOIN. DO causes the expr to be evaluated and the result discarded: the value of the whole loop is then 0. SUM causes all the values to be added and returned as the value of the loop, similarly PRODUCT causes them to be multiplied. COLLECT causes FOR to return a list containing all the values of expr. JOIN also causes FOR to return a list, consisting of all the values of exp (which must always be a list) spliced together (like APPEND in LISP).

REDUCE also allows
```
WHILE <boolean expression> DO <statement>
```
and its variant
```
REPEAT <statement> UNTIL <boolean expression>
```
The value is always zero.

R.2.4.5 *Blocks*

The syntax is quite simple:
```
BEGIN [<declaration>]* <statement>* END
```
However, various points must be noted. The only declarations which obey

the block-structure rules are simple variable declarations (see R.2.3.1): all others are global in scope. In particular, it is not possible to declare procedures within blocks. The value of a block is zero *unless* a RETURN <expression> has been executed, in which case the value of the expression becomes the value of the block.

It is also possible to have labels within blocks, introduced by the syntax <label>: before a statement. Labels can be any identifier (but not numbers). These labels can be used in GO TO (or GOTO) statements, *provided* that the GO TO is inside the same block as the label, and not inside any other blocks. In other words, GO TO cannot be used to leave or enter a block.

R.3 BUILT-IN FACILITIES

In the previous section we have discussed the syntax of REDUCE: here we describe the major algebraic facilities incorporated into REDUCE at the time this translation was prepared. Nevertheless, the definitive reference for this subject has to be the manual for the version of REDUCE available.

R.3.1 Prefix operators

R.3.1.1 *Numeric operators*

REDUCE understands the operators MIN and MAX, applied to an arbitrary number of numeric expressions. There is also ABS, which calculates the absolute value.

R.3.1.2 *Mathematical operators*

REDUCE has already defined the prefix operators COS, SIN, TAN, COT, ASIN, ACOS, ATAN, SQRT, EXP, LOG, SINH, COSH, TANH, ASINH, ACOSH, ATANH, ERF, DILOG and EXPINT. A few simplification rules are already known, and the user can define more with the LET statement (described in R.3.3.2). The switch setting on numval; causes instances of these operators with numeric arguments to be evaluated in floating point.

R.3.1.3 *Differentiation*

The operator DF performs differentiation. The first argument is the expression to be differentiated: subsequent expressions can be kernels (in which case they are the "variables" with respect to which differentiation is to be performed) or numbers (indicating how often the previous differentiation is to be performed). So df(a,x,2,y); calculates

$$\frac{\partial^3 a}{\partial x^2 \partial y}.$$

This takes account of dependencies (see the DEPEND statement in section 2.3.1), and any new rules introduced via LET.

```
1: operator f,g,h;

2: for all x,y let
2:      df(f(x,y),x)=g(x,y),
2:      df(f(x,y),y)=x*h(x,y);

3: df(f(a,b),a)*df(f(a,b),b);

G(A,B)*H(A,B)*A
```

R.3.1.4 *Integration*

The integration operator in REDUCE is INT, whose first argument is the integrand, and whose second is the variable of integration. The algorithm used is the "parallel" or Risch-Norman-Fitch method [Davenport, 1982]. We give a couple of examples (times are calculated on a SUN 3/75).

```
1: int(1/(1-x**4),x);

 2*ATAN(X) - LOG(X - 1) + LOG(X + 1)
-------------------------------------
                  4
Time: 357 ms


2: int(x**3*sin(x**2),x);

    2         2   2
 SIN(X ) - COS(X )*X
---------------------
          2

Time: 391 ms


3: int(1/(a*e**(m*x)+b*e**(-m*x)),x);

                           M*X
                         E    *A
 SQRT(B)*SQRT(A)*ATAN(-----------------)
                       SQRT(B)*SQRT(A)
----------------------------------------
                 A*B*M

Time: 2839 ms
```

R.3.1.5 *Factorisation*

REDUCE can factorise polynomials with integer coefficients, in one or several variables. There are two possible syntaxes:

`FACTORIZE(<expression>)`
which returns a list of all the factors (the first may be a numeric content), and
`FACTORIZE(<expression>,<prime>)`
which uses the given prime as a "hint" to the modular reduction process (see Chapter 4 for details of the algorithm used). Repeated factors appear the appropriate number of times in the list. The control option **on ifactor;** will also cause integer contents to be factorised as products of primes. It is also possible to factorise over numbers modulo a prime, and to get a description of the steps taken by the algorithm — see the manual for the details.

R.3.1.6 *Resultants*

REDUCE can compute the *resultant* of two expressions. The syntax is
`RESULTANT(<expression>,<expression>,<kernel>)`
where the third argument is the "indeterminate" to be eliminated.

R.3.1.7 *Solution of systems of equations*

The operator `SOLVE` can be used to solve a single polynomial equation, or a system of linear equations. The syntax is:
`SOLVE(<expression>,<variable>);`
where `<expression>` is either a single expression or a list of expressions, and `<variable>` is either a single variable, or a list of variables, representing the "unknowns" for which the system must be solved. If the second argument is omitted, then the equation(s) are solved for all the distinct kernels appearing.

```
1: solve(x**3+x**2+4=0,x);

        SQRT(7)*I - 1
{X= - ---------------,
            2

    SQRT(7)*I + 1
 X=---------------,
          2

 X=-2}

2: solve({x-3*y=2, x-y=1}, {x,y});

      1          1
{{X=---,Y= - ---}}
      2          2
```

There are various options that can be given — the reader should consult the REDUCE manual for the details.

R.3.2 Manipulation of expressions

First we should mention the concept of *current result*, which is denoted WS. This has, as its value, the last computed result. It is possible to access other results by using the notation WS <number>, which returns the result of the computation indexed by <number>.

R.3.2.1 *Output of expressions*

We have already mentioned that following statements with "$" rather than ";" prevents the result from being printed. Conversely, it is possible to cause a result to be printed with the WRITE statement. This takes a list of expressions, separated by commas, and prints the results.

The format of the output can be changed in many ways. The order in which the variables are printed can be changed by means of the ORDER statement.

```
order x,y,z;
```

orders x ahead of y ahead of z ahead of all other variables or kernels, in any future printing. The default order depends on the LISP system on which REDUCE is implemented.

```
1: p:=(x+y-z)**2;
```

$$P := X^2 + 2*X*Y - 2*X*Z + Y^2 - 2*Y*Z + Z^2$$

```
2: order z,y,x;

3: p;
```

$$Z^2 - 2*Z*Y - 2*Z*X + Y^2 + 2*Y*X + X^2$$

The FACTOR statement is similar, but causes all terms involving any particular power of the given kernels to be printed with that power factored out. For example, we can continue the previous example by:

```
4: factor z;

5: p;
```

$$Z^2 - (2*Z)*(Y + X) + Y^2 + 2*Y*X + X^2$$

The command REMFAC is used to cancel the effect of FACTOR declarations.

There are also many switches which control the output format of expressions. It should be noted that these do not affect the *internal* form, unlike switches such as MCD.

ALLFAC – prints simple multiplicative factors as such, and is normally on.

DIV – searches the denominator of rational expressions for simple factors, which are then divided into the numerator. This is off by

> default.
> LIST – prints each term in a sum on a separate line (good for using up paper!). This is normally off.
> RAT – is useful with FACTOR, and prints the denominator with each factored term (after cancellation if posible). It is normally off.
> RATPRI – prints fractions in the form $\frac{A}{B}$ if A and B each fit on one line. This is on by default.
> REVPRI – prints in ascending order, rather than descending order. It is normally off.
> NAT – invokes the "natural" or mathematical style of printing. It is normally on, but can be turned off to force printing in an output style which is compatible with the input syntax.
> FORT – causes expressions to be printed in FORTRAN-compatible format. There are several control parameters associated with this option — see the manual for the details.

It is often useful to be able to see the overall "structure" of a complex expression, or to see shared subparts as such. The STRUCTR statement displays this internal structure. For example,

```
1: (1+e^(log(x)*log log x))^5;

  5*LOG(LOG(X))*LOG(X)        4*LOG(LOG(X))*LOG(X)
E                      + 5*E                       + 10*

  3*LOG(LOG(X))*LOG(X)        2*LOG(LOG(X))*LOG(X)
E                      + 10*E                      + 5*

  LOG(LOG(X))*LOG(X)
E                    + 1


2: structr ws;

     5         4          3          2
ANS1   + 5*ANS1   + 10*ANS1   + 10*ANS1   + 5*ANS1 + 1

    where

                LOG(LOG(X))*LOG(X)
        ANS1 := E
```

This command can also be used with the FORT option, as in

```
3: on fort;

4: structr ws;
        ANS1=E**(LOG(LOG(X))*LOG(X))
        ANS=ANS1**5+5.*ANS1**4+10.*ANS1**3+10.*ANS1**2+5.*
       . ANS1+1.
```

Again, the manual should be consulted for the various options and additional possibilities.

R.3.2.2 *Parts of expressions*

The operator COEFF returns a list of the coefficients of a polynomial with respect to a named kernel.
```
1: coeff(x**3+(x-y)**2-1,x);
```

```
      2
{Y  - 1, - 2*Y,1,1}
```
It is possible to find just the n-th coefficient, by means of the operator COEFFN. This is more efficient than computing the whole list and then extracting one member.
```
2: coeffn((x+y)**3-2*x**2,x,2);
```

```
3*Y - 2
```


Various parts of rational functions and polynomials can be extracted with various operators:

NUM $-$ extracts the numerator of a rational function;

DEN $-$ extracts the denominator of a rational function;

DEG $-$ extracts the degree of a polynomial with respect to a particular kernel;

LCOF $-$ extracts the leading coefficient similarly;

LTERM $-$ extracts the leading term (coefficient times variable raised to the highest power);

REDUCT $-$ extracts the reductum (all except the leading term);

MAINVAR $-$ extracts the main variable of a polynomial.

The operator PART selects part of an expression. It works with respect to the *printed* form of the expression (bearing in mind all the switch settings described in the previous section on the output format). The syntax is
```
PART(<expression>[,<integer>]*).
```
The expression is notionally printed (in accordance with the printing control flags described in R.3.2.1), and then each <integer> n in turn is used to select the n-th argument of the current top-level operator in the expression. Negative integers select the $|n|$-th element from the end.

```
1: x:=a+b+c*d+e+f;
```

X := A + B + C*D + E + F

```
2: part(x,2);
```

B

```
3: part(x,-2);
```

E

```
4: part(x,3,1);
```

C

```
5: part(x,0);
```

PLUS

If we use **PART** on the left-hand side of an assignment statement, then the value of the assignment is the result of replacing the original part referred to by the right-hand side.

```
6: part(x,2):=g*h;
```

A + G*H + C*D + E + F

```
7: x;
```

A + B + C*D + E + F

The reader can see that the original expression is not changed.

R.3.3 Substitution

There are two kinds of substitution possible in REDUCE: *local* substitution commands, which take place in the command or expression being evaluated and *global* substitution commands, which affect all future calculations until they are cancelled.

R.3.3.1 *Local substitution*

This can be performed by the **SUB** operator, whose syntax is
SUB([<kernel>=<expressions>,]*<expressions>)
which yields the result of evaluating the last expression, and then replacing in parallel each kernel appearing as the left-hand side of an equation by the corresponding right-hand side. For example

```
1: sub(x=x+1, y=cos(u), x*y+u*sin(x));

SIN(X + 1)*U + COS(U)*X + COS(U)

2: z:=a*(x-y)**2+2*b*x+log(y);
```

$$Z := LOG(Y) + A*X^2 - 2*A*X*Y + A*Y^2 + 2*B*X$$

```
3: zp:=sub(x=y, y=u+1, z);
```

$$ZP := LOG(U + 1) + A*U^2 - 2*A*U*Y + 2*A*U + A*Y^2 - 2*A*Y + A +$$
$$2*B*Y$$

```
4: % But z is not altered
4: z;
```

$$LOG(Y) + A*X^2 - 2*A*X*Y + A*Y^2 + 2*B*X$$

A related operator is the infix operator WHERE, which has the syntax
<expression> WHERE <eqn>[,<eqn>]*.
In algebraic mode, WHERE and SUB provide the same capability, except that
WHERE can only substitute for identifiers, not for arbitrary kernels.

R.3.3.2 *Global substitution*

These substitutions, introduced by **LET** statements, take place in all expressions containing the left-hand side of the substitution rule, from the
moment when the rule is introduced until it is superceded, or cancelled via
the **CLEAR** command.

```
1: operator f,g;

2: let g(x)=x**2,
2:     f(x,y)=x*cos(y);

3: df(g(u),u); %our rule only affected g(x)

DF(G(U),U)

4: f(x,y);

COS(Y)*X

5: f(a,b); %our rule only affected f(x,y)

F(A,B)
```

If we wish to substitute for expressions containing arbitrary variables, rather
than particular named variables, then we must preface the **LET** statement
by **FOR ALL** preludes.

```
6: for all x,y let f(x,y)=log(x)-y;

7: f(a,b);
```

LOG(A) - B

The formal syntax is
`[FOR ALL <var>[,<var>]*] LET <rule>[,<rule>]*`
where a `rule` is defined as
`<lhs>=<expression>`
Various kinds of `lhs` are possible: the simplest being a kernel. In this case, the kernel is always replaced by the right-hand side. If the `lhs` is of the form `a**n` (where `n` is an explicit integer)*, then the rule applies to all powers of `a` of exponent at least `n`. If the `lhs` is of the form `a*b` (where `a` and `b` are kernels or powers of kernels or products of powers of kernels), then `a*b` will be replaced, but not `a` or `b` separately. If the `lhs` is of the form `a+b` (where `a` and `b` are kernels or powers of kernels or products of powers of kernels), then `a+b` will be replaced, and the *first* of `a` and `b` in REDUCE's ordering (say `a`) will be implicitly treated as `(a+b)-b`, and the `a+b` replaced.

It is also possible to place conditions, by means of `SUCH THAT` clauses, on the variables involved in a `FOR ALL` prelude, as in

```
1: operator laplace,factorial;

2: for all x,n such that fixp n and n>0 let
2:    laplace(x**n,x)=factorial(n)*t**(-n-1);

3: laplace(y**2,y);
```

```
 FACTORIAL(2)
 --------------
       3
       T
```

```
4: laplace(y**(-2),y);
```

```
          1
LAPLACE(----,Y)
          2
          Y
```

A rule can be cleared with the `CLEAR` command. To clear a rule en-

---

* One has to be more careful with other forms of exponents. For example, the rule `let a**b=c;` causes `a**(2*b);` to simplify to 0, since the internal form is essentially $\left(a^b\right)^2$, while `a**(b*d);` is not simplified, since its internal form is essentially a^{bd}.

tered with a prelude, it is necessary to use the *same* prelude, with the same variable names. We have already seen that CLEAR also clears WEIGHT declarations, and in fact it resets many other properties.

Unlike SUB and WHERE, LET rules are applied repeatedly until no further substitutions are possible. While this is often useful (e.g. in linearising trigonometric functions — see Section 2.7 in Chapter 2), it does mean that one should not reference the left-hand side of a rule in the right-hand side.

```
1: let x=x+1;
```

```
2: x;
```

***** Simplification recursion too deep

```
3: let y=z;
```

```
4: y;
```

Z

```
5: let z=y;
```

```
6: y;
```
***** Binding stack overflow, restarting...

R.4 MATRIX ALGEBRA

The MATRIX declaration lets one declare that variables will take matrix values, instead of the ordinary scalar ones. Matrices can be entered via the MAT operator, as in

```
1: matrix m;
```

```
2: m:=mat((a,b),(c,d));
```

M(1,1) := A

M(1,2) := B

M(2,1) := C

M(2,2) := D

REDUCE can compute the inverses of matrices, their determinants, transposes and traces, as in

```
3: m**(-1);
```

$$MAT(1,1) := \frac{D}{A*D - B*C}$$

$$MAT(1,2) := - \frac{B}{A*D - B*C}$$

$$MAT(2,1) := - \frac{C}{A*D - B*C}$$

$$MAT(2,2) := \frac{A}{A*D - B*C}$$

```
4: det(m);
```

```
A*D - B*C
```

```
5: tp(m);
```

```
MAT(1,1) := A
```

```
MAT(1,2) := C
```

```
MAT(2,1) := B
```

```
MAT(2,2) := D
```

```
6: trace m;
```

```
A + D
```

R.5 CONCLUSION

REDUCE is a powerful Computer Algebra system available on many machines, from the IBM PC to the largest main-frames, but the elementary facilities are easy to use. Full details (and many options and variations that we have not had the space to present here) are contained in the manual.

Bibliography

This book has only given a brief outline of Computer Algebra. Before giving a list of all the work referred to in this book, we shall mention several books and periodicals of general interest, which may add to the reader's knowledge of this subject. The mathematics and algorithms for treating integers and dense polynomials have been very carefully dealt with by Knuth [1981]. The collection edited by Buchberger *et al.* [1982] describes several mathematical aspects of Computer Algebra, and states many of the proofs we have omitted. These two books are essential for those who want to create their own system.

Several articles on Computer Algebra and its applications have appeared in the proceedings of conferences on Computer Algebra. The conferences in North America are organised by **SIGSAM** (the Special Interest Group on Symbolic and Algebraic Computation of the Association for Computing Machinery) and their proceedings are published by the ACM, with the title SYMSAC, such as SYMSAC 76 (Yorktown Heights), SYMSAC 81 (Snowbird) and SYMSAC 86 (Waterloo). The proceedings of the European conferences have been published since 1979 by Springer-Verlag in their series *Lecture Notes in Computer Science*: The following table gives the details.

Conference	Place	LNCS
EUROSAM 79	Marseilles	72
EUROCAM 82	Marseilles	144
EUROCAL 83	Kingston-on-Thames	162
EUROSAM 84	Cambridge	174
EUROCAL 85	Linz	203 and 204

For each system of Computer Algebra the best reference book for the user is, of course, its manual. But these manuals are often quite long, and it is hard to get an overall view of the possibilities and limitations of these systems. Thus, the *Journal of Symbolic Computation* (published by Academic Press) contains descriptions of systems, presentations of new applications of Computer Algebra, as well as research articles in this field.

The other journal of particular interest to students of Computer Algebra is the *SIGSAM Bulletin*, published by SIGSAM, which we have already mentioned. It is a very informal journal, which contains, as well as scientific articles, problems, announcements of conferences, research reports, new versions of systems etc.

In France, there is a GRECO of Computer Algebra, which organises a conference every year. The proceedings are published along with other interesting articles in the review *CALSYF*, edited by M. Mignotte at the University of Strasbourg. With this one can follow French work in this field.

[Abbott *et al.*, 1985] Abbott,J.A., Bradford,R.J. & Davenport,J.H., A Remark on Factorisation. SIGSAM Bulletin **19** (1985) 2, pp. 31–33, 37.

[Abbott *et al.*, 1987] Abbott,J.A., Bradford,R.J. & Davenport,J.II., A Remark on Sparse Polynomial Multiplication. To appear.

[Abdali *et al.*, 1977] Abdali,S.K., Caviness,B.F. & Pridor,A., Modular Polynomial Arithmetic in Partial Fraction Decomposition. Proc. 1977 MACSYMA Users' Conference, NASA publication CP–2012, National Technical Information Service, Springfield, Virginia., pp. 253–261.

[Aho *et al.*, 1974] Aho,A.V., Hopcroft,J.E. & Ullman,J.D., The Design and Analysis of Computer Algorithms. Addison-Wesley, 1974. MR **54** (1977) #1706.

[Allen, 1978] Allen,J.R., The Anatomy of LISP. McGraw-IIill, New York, 1978.

[Arnon, 1985] Arnon,D.S., On Mechanical Quantifier Elimination for Elementary Algebra and Geometry: Solution of a Nontrivial Problem. Proc. EUROCAL 85, Vol. 2 [Springer Lecture Notes in Computer Science Vol. 204, Springer-Verlag, 1985] pp. 270–271.

[Arnon & Smith, 1983] Arnon,D.S. & Smith,S.F., Towards Mechanical Solution of the Kahan Ellipse Problem I. Proc. EUROCAL 83 [Springer Lecture Notes in Computer Science 162, Springer-Verlag, Berlin, Heidelberg, New York, 1983], pp. 36–44.

[Arnon *et al.*, 1984a] Arnon,D.S, Collins,G.E. & McCallum,S., Cylindrical Algebraic Decomposition I: The Basic Algorithm. SIAM J. Comp.

13 (1984) pp. 865–877.

[Arnon *et al.*, 1984b] Arnon,D.S, Collins,G.E. & McCallum,S., Cylindrical Algebraic Decomposition I: An Adjacency Algorithm for the Plane. SIAM J. Comp. **13** (1984) pp. 878–889.

[Bareiss, 1968] Bareiss,E.H., Sylvester's Identity and Multistep Integer-preserving Gaussian Elimination. Math. Comp. **22** (1968) pp. 565–578. Zbl. 187,97.

[Bateman & Danielopoulos, 1981] Bateman,S.O.,Jr., & Danielopoulos,S.D., Computerised Analytic Solutions of Second Order Differential Equations. Computer J. **24** (1981) pp. 180–183. Zbl. 456.68043. MR 82m:68075.

[Berlekamp, 1967] Berlekamp,E.R., Factoring Polynomials over Finite Fields. Bell System Tech. J. **46** (1967) pp. 1853–1859.

[Borodin *et al.*, 1985] Borodin,A., Fagin,R., Hopcroft,J.E. & Tompa,M., Decreasing the Nesting Depth of an Expression Involving Square Roots. J. Symbolic Comp. **1** (1985), pp. 169–188.

[Brent, 1970] Brent,R.P., Algorithms for Matrix Multiplication. Report CS 157, Computer Science Department, Stanford University, March 1970. (The results are described by Knuth [1981], p. 482.)

[Brown, 1969] Brown,W.S., Rational Exponential Expressions and a Conjecture concerning π and e. Amer. Math. Monthly **76** (1969) pp. 28–34.

[Brown, 1971] Brown,W.S., On Euclid's Algorithm and the Computation of Polynomial Greatest Common Divisors. J. ACM **18** (1971) pp. 478–504. MR **46** (1973) #6570.

[Buchberger, 1970] Buchberger,B., Ein algorithmisches Kriterium für die Lösbarkeit eines algebraischen Gleichungssystems. Aequationes Mathematicæ **4** (1970) pp. 374–383. (An algorithmic criterion for the solubility of an algebraic system of equations.)

[Buchberger, 1976a] Buchberger,B., Theoretical Basis for the Reduction of Polynomials to Canonical Forms. SIGSAM Bulletin **39** (Aug. 1976) pp. 19–29.

[Buchberger, 1976b] Buchberger,B., Some Properties of Gröbner-Bases for Polynomial Ideals. SIGSAM Bulletin **40** (Nov. 1976) pp. 19–24.

[Buchberger, 1979] Buchberger,B., A Criterion for Detecting Unnecessary Reductions in the Construction of Groebner Bases. Proceedings of the 1979 European Symposium on Symbolic and Algebraic Computation [Springer Lecture Notes in Computer Science 72, Springer-Verlag, Berlin, Heidelberg, New York, 1979], pp. 3–21. Zbl. 417.68029. MR 82e:14004.

[Buchberger, 1981] Buchberger,B., H-Bases and Gröbner-Bases for Polynomial Ideals. CAMP-Linz publication 81–2.0, University of Linz, Feb. 1981.

[Buchberger, 1983] Buchberger,B., A Note on the Complexity of Computing Gröbner-Bases. Proc. EUROCAL 83 [Springer Lecture Notes in Computer Science 162, Springer-Verlag, Berlin, Heidelberg, New York, 1983], pp. 137–145.

[Buchberger, 1985] Buchberger,B., A Survey on the Method of Groebner bases for Solving Problems in Connection with Systems of Multivariate Polynomials. Proc. 2nd RIKEN Symposium Symbolic & Algebraic Computation (ed. N. Inada & T. Soma), World Scientific Publ., 1985, pp. 69–83.

[Buchberger *et al.*, 1982] Buchberger,B., Collins,G.E. & Loos,R. (editors), Symbolic & Algebraic Computation. Computing Supplementum 4, Springer-Verlag, Wien, New York, 1982.

[Capelli, 1901] Capelli,A., Sulla riduttibilita della funzione $x^n - A$ in campo qualunque di rationalità. Math. Ann. **54** (1901) pp. 602–603. (On the reducibility of the function $x^n - A$ in some rational field.)

[Cauchy, 1829] Cauchy,A.-L., Exercises de Mathématiques Quatrième Année. De Bure Frères, Paris, 1829. Œuvres, Sér. II, Vol. IX, Gauthier-Villars, Paris, 1891.

[Char *et al.*, 1985] Char,B.W., Geddes,K.O., Gonnet,G.H., Watt,S.M., Maple user's guide. Watcom Publications, Waterloo 1985.

[Cherry, 1983] Cherry,G.W., Algorithms for Integrating Elementary Functions in Terms of Logarithmic Integrals and Error Functions. Ph.D. Thesis, Univ. Delaware, August 1983.

[Cherry, 1985] Cherry,G.W., Integration in Finite Terms with Special Functions: the Error Function. J. Symbolic Comp. **1** (1985) pp. 283–302.

[Cherry & Caviness, 1984] Cherry,G.W. & Caviness,B.F., Integration in Finite Terms with Special Functions: A Progress Report. Proc. EUROSAM 84 [Springer Lecture Notes in Computer Science 174, Springer-Verlag, Berlin, Heidelberg, New York, Tokyo, 1984], pp. 351–359.

[Chou & Collins, 1982] Chou,T.-W.J. & Collins,G.E., Algorithms for the solution of systems of linear Diophantine equations. SIAM J. Comp. **11** (1982) pp. 687–708. Zbl. 498.65022. MR 84e:10020.

[Collins, 1971] Collins,G.E., The Calculation of Multivariate Polynomial Resultants. J. ACM **18** (1971) pp. 515–532.

[Collins,1975] Collins,G.E., Quantifier Elimination for Real Closed Fields by Cylindrical Algebraic Decomposition. Proc. 2nd GI Conf. Automata Theory and Formal Languages (Springer Lecture Notes in Computer Science 33), pp. 134–183. MR **55** (1977) #771.

[Collins & Loos, 1982] Collins,G.E. & Loos,R., Real Zeros of Polynomials. Symbolic & Algebraic Computation (Computing Supplementum 4) (ed. B. Buchberger, G.E. Collins & R. Loos) Springer-Verlag, Wien, New York, 1982, pp. 83–94.

[Coppersmith & Davenport, 1985] Coppersmith,D. & Davenport,J.H., An Application of Factoring. J. Symbolic Comp. **1** (1985) pp. 241–243.

[Coppersmith & Winograd, 1982] Coppersmith,D. & Winograd,S., On the Asymptotic Complexity of Matrix Multiplication. SIAM J. Comp. **11** (1982) pp. 472–492. Zbl. 486.68030. MR 83j:68047b.

[Coppersmith *et al.*, 1986] Coppersmith,D., Odlyzko,A.M. & Schroeppel,R., Discrete Logarithms in $GF(p)$. Algorithmica **1** (1986) pp. 1–15.

[Coxeter, 1961] Coxeter,H.S.M., Introduction to Geometry. Wiley, New York, 1961.

[Davenport, 1981] Davenport,J.H., On the Integration of Algebraic Functions. Springer Lecture Notes in Computer Science 102, Springer-Verlag, Berlin, Heidelberg, New York, 1981. Zbl. 471.14009. MR 84k:14024.

[Davenport,1982] Davenport,J.H., On the Parallel Risch Algorithm (I). Proc. EUROCAM '82 [Springer Lecture Notes in Computer Science 144, Springer-Verlag, Berlin, Heidelberg, New York, 1982], pp. 144–157. MR 84b:12033.

[Davenport, 1984a] Davenport,J.H., Intégration algorithmique des fonctions élémentairement transcendantes sur une courbe algébrique. Annales de l'Institut Fourier **34** (1984) pp. 271–276. Zbl. 506.34002.

[Davenport, 1984b] Davenport,J.H., $y' + fy = g$. Proc. EUROSAM 84 [Springer Lecture Notes in Computer Science 174, Springer-Verlag, Berlin, Heidelberg, New York, Tokyo, 1984], pp. 341–350.

[Davenport, 1984c] Davenport,J.H., Solutions of Inhomogenous Differential Equations. preprint, Journées équations différentielles dans le champ complexe, June 1984. To appear in the SIGSAM Bulletin.

[Davenport, 1985a] Davenport,J.H., Closed Form Solutions of Ordinary Differential Equations. Proc. 2nd RIKEN Symposium Symbolic & Algebraic Computation (ed. N. Inada & T. Soma), World Scientific Publ., 1985, pp. 183–195.

[Davenport, 1985b] Davenport,J.H., Computer Algebra for Cylindrical Algebraic Decomposition. TRITA–NA–8511, NADA, KTH, Stockholm, Sept. 1985.

[Davenport, 1985c] Davenport,J.H., On the Risch Differential Equation Problem. SIAM J. Comp. **15** (1986) pp. 903–918.

[Davenport, 1986] Davenport,J.H., On a "Piano Movers" Problem. SIGSAM Bulletin **20** (1986) 1&2 pp. 15–17.

[Davenport & Heintz, 1987] Davenport,J.H. & Heintz,J., Real Quantifier Elimination is Doubly Exponential. To appear in J. Symbolic Comp.

[Davenport & Padget 1987] Davenport,J.H. & Padget,J.A., On Number Bases for Symbolic Computation. To appear.

[Davenport & Tournier, 1983] Davenport,J.H. & Tournier,E., Définition syntaxique de RLISP. Rapport Interne, IMAG, 1983.

[Delaunay, 1860] Delaunay, Ch., Théorie du mouvement de la lune. Extract from the Comptes Rendus de l'Académie des Sciences, Vol. LI.

[Della Dora & Tournier, 1981] Della Dora,J. & Tournier,E., Formal Solutions of Differential Equations in the Neighbourhood of Singular Points (Regular and Irregular). Proceedings of the 1981 ACM Symposium on Symbolic and Algebraic Computation, ACM Inc., New York, 1981, pp. 25–29.

[Della Dora & Tournier, 1984] Della Dora,J. & Tournier,E., Homogeneous Linear Differential Equations (Frobenius-Boole Method). Proc. EUROSAM 84 [Springer Lecture Notes in Computer Science 174, Springer-Verlag, Berlin, Heidelberg, New York, Tokyo, 1984], pp. 1–12.

[Della Dora & Tournier, 1986] Della Dora,J. & Tournier,E., Solutions formelles asymptotiques d'équations récurrentes linéaires: méthode de Pincherle-Ramis. Research report, IMAG, Grenoble.

[Della Dora *et al.*, 1982] Della Dora,J., Dicrescenzo,C. & Tournier,E., An Algorithm to Obtain Formal Solutions of a Linear Homogeneous Differential Equation at an Irregular Singular Point. Proc. EUROCAM 82 [Springer Lecture Notes in Computer Science 144, Springer-Verlag, Berlin, Heidelberg, New York, 1982], pp. 273–280. MR 84c:65094.

[Della Dora *et al.*, 1985] Della Dora,J., Dicrescenzo,C. & Duval,D., About a new Method for Computing in Algebraic Number Fields. Proc. EUROCAL 85, Vol. 2 [Springer Lecture Notes in Computer Science 204, Springer-Verlag,Berlin, Heidelberg, New York, Tokyo, 1985], pp. 289–290.

[Dicrescenzo & Duval, 1984] Dicrescenzo,C. & Duval,D., Computations on Curves. Proc. EUROSAM 84 [Springer Lecture Notes in Computer Science 174, Springer-Verlag, Berlin, Heidelberg, New York, Tokyo, 1984], pp. 100–107.

[Dicrescenzo & Duval, 1985] Dicrescenzo,C. & Duval,D., Algebraic Computation on Algebraic Numbers. Computers and Computing (ed. P. Chenin, C. Dicrescenzo, F. Robert), Masson and Wiley, 1985, pp. 54–61.

[Duval, 1987a] Duval,D., An algorithmic proof of the resultant formula. To appear (also in [Duval, 1987b]).

[Duval, 1987b] Duval,D., Diverses questions relatives au Calcul Formel avec des nombres algébriques. Thèse d'État, Université I de Grenoble, April 1987.

[Dubreuil, 1963] Dubreuil,P., Algèbre. Gauthier-Villars, Paris, 1963.

[Ehrhard, 1986] Ehrhard,T., Personal communication, March 1986.

[Fitch, 1974] Fitch,J.P., CAMAL Users' Manual. University of Cambridge Computer Laboratory, 1974.

[Fitch, 1985] Fitch,J.P., Solving Algebraic Problems with REDUCE. J. Symbolic Comp. **1** (1985), pp. 211–227.

[Gebauer & Kredel, 1984] Gebauer,P. & Kredel,H., Note on "Solution of a General System of Equations". SIGSAM Bulletin **18** (1984) 3, pp. 5–6.

[Giusti, 1984] Giusti,M., Some Effectivity Problems in Polynomial Ideal Theory. Proc. EUROSAM 84 [Springer Lecture Notes in Computer Science 174, Springer-Verlag, Berlin, Heidelberg, New York, Tokyo, 1984], pp. 159–171.

[Gregory & Krishnamurthy, 1984] Gregory,R.T. & Krishnamurthy,E.V., Methods and Applications of Error-free Computation. Springer-Verlag, New York, 1984. CR 8504–0270.

[Griesmer *et al.*, 1975] Griesmer,J.H., Jenks,R.D. & Yun,D.Y.Y., SCRATCHPAD User's Manual. IBM Research Publication RA70, June 1975.

[Hadamard, 1893] Hadamard,J., Résolution d'une Question Relative aux Déterminants. Bull. des Sci. Math. (2) **17** (1893) pp. 240–246. Œuvres, CNRS, Paris, 1968, Vol. I, pp. 239–245.

[Hearn, 1987] Hearn,A.C., REDUCE-3 User's Manual, version 3.3, Rand Corporation Publication CP78 (7/87), 1987.

[Heindel,1971] Heindel,L.E., Integer Arithmetic Algorithm for Polynomial Real Zero Determination. J. ACM **18** (1971) pp. 535–548.

[Hermite, 1872] Hermite,C., Sur l'intégration des fractions rationelles. Nouvelles Annales de Mathématiques, 2 Sér., **11** (1872) pp. 145–148. Ann. Scientifiques de l'Ecole Normale Supérieure, 2 Sér., **1** (1872) pp. 215–218.

[Hilali, 1982] Hilali,A., Contribution à l'étude de points singuliers des systèmes différentiels linéaires. Thèse de troisième cycle, IMAG, Grenoble, 1982.

[Hilali, 1983] Hilali,A., Characterization of a Linear Differential System with a Regular Singularity. Proc. EUROCAL 83 [Springer Lecture Notes in Computer Science 162, Springer-Verlag, Berlin, Heidelberg, New York, 1983], pp. 68–77.

[Hilali, 1987] Hilali,A., Solutions formelles de systèmes différentiels linéaires au voisinage de points singuliers. Thèse d'État, Université I de Grenoble, June 1987.

[Horowitz, 1969] Horowitz,E., Algorithm for Symbolic Integration of Rational Functions. Ph.D. Thesis, Univ. of Wisconsin, November 1969.

[Horowitz, 1971] Horowitz,E., Algorithms for Partial Fraction Decomposition and Rational Function Integration. Proc. Second Symposium on Symbolic and Algebraic Manipulation, ACM Inc., 1971, pp. 441–457.

[Ince, 1953] Ince,F.L., Ordinary Differential Equations. Dover Publications, 1953.

[Jenks, 1984] Jenks,R.D., A Primer: 11 Keys to New SCRATCHPAD. Proc. EUROSAM 84 [Springer Lecture Notes in Computer Science 174, Springer-Verlag, Berlin, Heidelberg, New York, Tokyo, 1984], pp. 123–147.

[Johnson, 1974] Johnson,S.C., Sparse Polynomial Arithmetic. Proc EUROSAM 74 (SIGSAM Bulletin Vol. 8, No. 3, Aug. 1974, pp. 63–71).

[Kahrimanian, 1953] Kahrimanian,H.G., Analytic differentiation by a digital computer. M.A. Thesis, Temple U., Philadelphia, Pennsylvania, May 1953.

[Kaltofen *et al.*, 1981] Kaltofen,E., Musser,D.R. & Saunders,B.D., A Generalized Class of Polynomials that are Hard to Factor. Proceedings of the 1981 ACM Symposium on Symbolic and Algebraic Computation, ACM Inc., New York, 1981, pp. 188–194. Zbl. 477.68041.

[Kaltofen *et al.*, 1983] Kaltofen,E., Musser,D.R. & Saunders,B.D., A Generalized Class of Polynomials that are Hard to Factor. SIAM J.

Comp. **12** (1983) pp. 473–483. CR 8504-0367 (Vol. **25** (1985) p. 235). MR 85a:12001.

[Knuth, 1969] Knuth,D.E., The Art of Computer Programming, Vol. II, Semi-numerical Algorithms, Addison-Wesley, 1969.

[Knuth, 1973] Knuth,D.E., The Art of Computer Programming, Vol. I, Fundamental Algorithms, 2nd Edn., Addison-Wesley, 1973.

[Knuth, 1981] Knuth,D.E., The Art of Computer Programming, Vol. II, Semi-numerical Algorithms, 2nd Edn., Addison-Wesley, 1981. MR 83i:68003.

[Kovacic, 1977] Kovacic,J.J., An Algorithm for solving Second Order Linear Homogeneous Differential Equations. Preprint, Brooklyn College, City University of New York.

[Kovacic, 1986] Kovacic,J.J., An Algorithm for solving Second Order Linear Homogeneous Differential Equations. J. Symbolic Comp. **2** (1986) pp. 3–43.

[Krishnamurthy, 1985] Krishnamurthy,E.V., Error-free Polynomial Matrix Computations. Springer-Verlag, New York, 1985.

[Landau, 1905] Landau,E., Sur Quelques Théorèmes de M. Petrovitch Relatifs aux Zéros des Fonctions Analytiques. Bull. Soc. Math. France **33** (1905) pp. 251–261.

[Lang, 1965] Lang,S., Algebra. Addison-Wesley, Reading, Mass., 1959.

[Lauer, 1982] Lauer,M., Computing by Homomorphic Images. Symbolic & Algebraic Computation (Computing Supplementum 4) (ed. B. Buchberger, G.E. Collins & R. Loos) Springer-Verlag, Wien, New York, 1982, pp. 139–168.

[Lauer, 1983] Lauer,M., Generalized *p*-adic Constructions. SIAM J. Comp. **12** (1983) pp. 395–410. Zbl. 513.68035.

[Lazard, 1983] Lazard,D., Gröbner Bases, Gaussian Elimination and Resolution of Systems of Algebraic Equations. Proc. EUROCAL 83 [Springer Lecture Notes in Computer Science 162, Springer-Verlag, Berlin, Heidelberg, New York, 1983], pp. 146–157.

[Lazard, 1987] Lazard,D., Quantifier Elimination: Optimal Solution for 2 Classical Examples. To appear in J. Symbolic Comp.

[Lenstra *et al.*, 1982] Lenstra,A.K., Lenstra,H.W.,Jr. & Lovász,L., Factoring Polynomials with Rational Coefficients. Math. Ann. **261** (1982) pp. 515–534. Zbl. 488.12001. MR 84a:12002.

[Lipson, 1976] Lipson,J.D., Newton's Method: a great Algebraic Algorithm. Proceedings of the 1976 ACM Symposium on Symbolic and Algebraic Computation, ACM Inc., New York, 1976, pp. 260–270. Zbl. 454.65035.

[Loos, 1982] Loos,R., Generalized Polynomial Remainder Sequences. Symbolic & Algebraic Computation (Computing Supplementum 4) (ed. B. Buchberger, G.E. Collins & R. Loos), Springer-Verlag, Wien, New York, 1982, pp. 115–137.

[McCallum, 1985a] McCallum,S., An Improved Projection Algorithm for Cylindrical Algebraic Decomposition. Computer Science Tech. Report 548, Univ. Wisconsin at Madison, 1985.

[McCallum, 1985b] McCallum,S., An Improved Projection Algorithm for Cylindrical Algebraic Decomposition. Proc. EUROCAL 85, Vol. 2 [Springer Lecture Notes in Computer Science 204, Springer-Verlag, Berlin, Heidelberg, New York, Tokyo, 1985], pp. 277–278.

[McCarthy et al., 1965] McCarthy,J., Abrahams,P.W., Edwards,W., Hart, T.P. & Levin,M., The LISP 1.5 Programmers Manual. M.I.T. Press, 1965.

[Macmillan & Davenport, 1984] Macmillan,R.J. & Davenport,J.H., Factoring Medium-Sized Integers. Computer J. **27** (1984) pp. 83–84.

[Marti et al., 1978] Marti,J.B., Hearn,A.C., Griss,M.L. & Griss,C., The Standard LISP Report. Report UCP-60, University of Utah, Jan. 1978. SIGSAM Bulletin **14** (1980), 1, pp. 23–43.

[Mauny, 1985] Mauny,M., Thèse de troisième cycle, Paris VII, Sept. 1985.

[Mayr & Mayer, 1982] Mayr,E. & Mayer,A., The Complexity of the Word Problem for Commutative Semi-groups and Polynomial Ideals. Adv. in Math. **46** (1982) pp. 305–329.

[Mignotte, 1974] Mignotte,M., An Inequality about Factors of Polynomials. Math. Comp. **28** (1974) pp. 1153–1157. Zbl. 299.12101.

[Mignotte, 1981] Mignotte,M., Some Inequalities About Univariate Polynomials. Proceedings of the 1981 ACM Symposium on Symbolic and Algebraic Computation, ACM Inc., New York, 1981, pp. 195–199. Zbl. 477.68037.

[Mignotte, 1982] Mignotte,M., Some Useful Bounds. Symbolic & Algebraic Computation (Computing Supplementum 4) (ed. B. Buchberger, G.E. Collins & R. Loos), Springer-Verlag, Wien, New York, 1982, pp. 259–263. Zbl. 498.12019.

[Mignotte, 1986] Mignotte,M., Computer versus Paper and Pencil. CAL-SYF **4**, pp. 63–69.

[Moses, 1966] Moses,J., Solution of a System of Polynomial Equations by Elimination. Comm. ACM **9** (1966) pp. 634–637.

[Moses, 1967] Moses,J., Symbolic Integration. Ph.D. Thesis & Project MAC TR–47, M.I.T., 1967.

[Moses, 1971a] Moses,J., Algebraic Simplification — A Guide for the Perplexed. Comm. ACM **14** (1971) pp. 527–537.

[Moses, 1971b] Moses,J., Symbolic Integration, the stormy decade. Comm. ACM **14** (1971) pp. 548–560.

[Musser, 1978] Musser,D.R., On the Efficiency of a Polynomial Irreducibility Test. J. ACM **25** (1978) pp. 271–282. MR 80m:68040.

[Najid-Zejli, 1984] Najid-Zejli,H., Computation in Radical Extensions. Proc. EUROSAM 84 [Springer Lecture Notes in Computer Science 174, Springer-Verlag, Berlin, Heidelberg, New York, Tokyo, 1984], pp. 115–122.

[Najid-Zejli, 1985] Najid-Zejli,H., Extensions algébriques: cas général et cas des radicaux. Thèse de troisième cycle, IMAG, Grenoble, 25.6.85.

[Nolan, 1953] Nolan,J., Analytic differentiation on a digital computer. M.A. Thesis, Math. Dept., M.I.T., Cambridge, Massachusetts, May 1953.

[Norman, 1975] Norman,A.C., Computing with Formal Power Series. ACM Transactions on Mathematical Software **1** (1975) pp. 346–356. Zbl. 315.65044.

[Norman, 1982] Norman,A.C., The Development of a Vector-based Algebra System. Proc. EUROCAM 82 [Springer Lecture Notes in Computer Science 144, Springer-Verlag, Berlin, Heidelberg, New York, 1982], pp. 237–248.

[Norman & Davenport, 1979] Norman,A.C., & Davenport,J.H., Symbolic Integration — the Dust Settles? Proceedings of the 1979 European Symposium on Symbolic and Algebraic Computation [Springer Lecture Notes in Computer Science 72, Springer-Verlag, Berlin, Heidelberg, New York 1979], pp. 398–407. Zbl. 399.68056. MR 82b:68031.

[Norman & Moore, 1977] Norman,A.C., & Moore,P.M.A., Implementing the new Risch Integration Algorithm. Proc. Symp. advanced computing methods in theoretical physics, Marseilles, 1977, pp. 99–110.

[Ostrowski, 1946] Ostrowski,A.M., Sur l'intégrabilité élémentaire de quelques classes d'expressions. Comm. Math. Helvet. **18** (1946) pp. 283–308.

[Pearce & Hicks, 1981] Pearce,P.D. & Hicks,R.J., The Optimization of User Programs for an Algebraic Manipulation System. Proceedings of the 1981 ACM Symposium on Symbolic and Algebraic Computation, ACM Inc., New York, 1981, pp. 131–136.

[Pearce & Hicks, 1982] Pearce,P.D. & Hicks,R.J., The Application of Algebraic Optimisation Techniques to Algebraic Mode Programs for REDUCE. SIGSAM Bulletin **15** (1981/2) 4, pp. 15–22.

[Pearce & Hicks, 1983] Pearce,P.D. & Hicks,R.J., Data Structures and Execution Times of Algebraic Mode Programs for REDUCE. SIGSAM Bulletin **17** (1983) 1, pp. 31–37.

[Probst & Alagar, 1982] Probst,D. & Alagar,V.S., An Adaptive Hybrid Algorithm for Multiplying Dense Polynomials. Proc. EUROCAM 82 [Springer Lecture Notes in Computer Science 144, Springer-Verlag, Berlin, Heidelberg, New York, 1982], pp. 16–23.

[Puiseux, 1850] Puiseux,M.V., Recherches sur les fonctions algébriques. J. Math. Pures et Appliquées **15** (1850) pp. 365-480.

[Ramanujan, 1927] Ramanujan,S., Problems and Solutions. In Collected Works (ed. G.H. Hardy, P.V. Secha Ayar, & B.M. Wilson), C.U.P., 1927.

[Ramis & Thomann, 1980] Ramis,J.P. & Thomann,J., Remarques sur l'utilisation numérique des séries de factorielles. Séminaire d'analyse numérique, Strasbourg No. 364, 1980.

[Ramis & Thomann, 1981] Ramis,J.P. & Thomann,J., Some Comments about the Numerical Utilization of Factorial Series Methods in the Study of Critical Phenomena. Springer-Verlag, Berlin, Heidelberg, New York, 1981.

[Richard, 1986] Richard,F., Représentations graphiques de solutions d'équations différentielles dans le champ complexe. Rapport de recherche, IRMA, Strasbourg.

[Richards & Whitby-Strevens, 1979] Richards,M. & Whitby-Strevens,C., BCPL, The Language and its Compiler. C.U.P., 1979. Zbl. 467.68004.

[Richardson, 1968] Richardson,D., Some Unsolvable Problems Involving Elementary Functions of a Real Variable. J. Symbolic Logic **33** (1968), pp. 511–520.

[Risch, 1969] Risch,R.H., The Problem of Integration in Finite Terms. Trans. A.M.S. **139** (1969) pp. 167–189. Zbl. 184,67. MR **38** (1969) #5759.

[Risch, 1979] Risch,R.H., Algebraic Properties of the Elementary Functions of Analysis. Amer. J. Math. **101** (1979) pp. 743–759. MR 81b:12029.

[Rosenlicht, 1976] Rosenlicht,M., On Liouville's Theory of Elementary Functions. Pacific J. Math **65** (1976), pp. 485–492.

[Rothstein, 1976] Rothstein,M., Aspects of Symbolic Integration and Simplification of Exponential and Primitive Functions. Ph.D. Thesis, Univ. of Wisconsin, Madison, 1976. (Xerox University Microfilms 77–8809.)

[Sasaki & Murao, 1981] Sasaki,T. & Murao,H., Efficient Gaussian Elimination Method for Symbolic Determinants and Linear Systems. Proceedings of the 1981 ACM Symposium on Symbolic and Algebraic Computation, ACM Inc., New York, 1981, pp. 155–159. Zbl. 486.68023.

[Sasaki & Murao, 1982] Sasaki,T. & Murao,H., Efficient Gaussian Elimination Method for Symbolic Determinants and Linear Systems. ACM Transactions on Mathematical Software 8 (1982) pp. 277–289. Zbl. 491.65014. CR 40, 106 (Vol. **24** (1983) p. 103).

[Saunders, 1981] Saunders,B.D., An Implementation of Kovacic's Algorithm for Solving Second Order Linear Homogeneous Differential Equations. Proceedings of the 1981 ACM Symposium on Symbolic and Algebraic Computation, ACM Inc., New York, 1981, pp. 105–108. Zbl. 486.68023.

[Schwartz & Sharir, 1983a] Schwartz,J.T. & Sharir,M., On the "Piano Movers" Problem II. General Techniques for Computing Topological Properties of Real Algebraic Manifolds. Advances Appl. Math. 4 (1983) pp. 298–351.

[Schwartz & Sharir, 1983b] Schwartz,J.T. & Sharir,M., On the "Piano Movers" Problem II. Coordinating the Motion of Several Independent Bodies: The Special Case of Circular Bodies Moving Amidst Polygonal Barriers. Int. J. Robot. Res. 2 (1983) pp. 46–75.

[Singer, 1981] Singer,M.F., Liouvillian Solutions of n-th Order Homogeneous Linear Differential Equations. Amer. J. Math. **103** (1981) pp. 661–682. Zbl. 477.12016. MR 82i:12008.

[Singer, 1985] Singer,M.F., Solving Homogeneous Linear Differential Equations in Terms of Second Order Linear Differential Equations. Amer. J. Math. **107** (1985) pp. 663–696.

[Singer & Davenport, 1985] Singer,M.F. & Davenport,J.H., Elementary and Liouvillian Solutions of Linear Differential Equations. Proc. EUROCAL 85, Vol. 2 [Springer Lecture Notes in Computer Science 204, Springer-Verlag, Berlin, Heidelberg, New York, Tokyo, 1985], pp. 595–596.

[Singer & Davenport, 1986] Singer,M.F. & Davenport,J.H., Elementary and Liouvillian Solutions of Linear Differential Equations. J. Symbolic Comp. 2 (1986) pp. 237–260.

[Singer *et al.*, 1981] Singer,M.F., Saunders,B.D. & Caviness,B.F., An Extension of Liouville's Theorem on Integration in Finite Terms. Proceedings of the 1981 ACM Symposium on Symbolic and Algebraic Computation, ACM Inc., New York, 1981, pp. 23–24. Zbl. 482.12008.

[Singer *et al.*, 1985] Singer,M.F., Saunders,B.D. & Caviness,B.F., An Extension of Liouville's Theorem on Integration in Finite Terms. SIAM J. Comp. **14** (1985) pp. 966–990.

[Slagle, 1961] Slagle,J., A Heuristic Program that Solves Symbolic Integration Problems in Freshman Calculus. Ph.D. Dissertation, Harvard U., Cambridge, Mass. May 1961.

[Smit, 1981] Smit,J., A Cancellation Free Algorithm, with Factoring Capabilities, for the Efficient Solution of Large Sparse Sets of Equations. Proceedings of the 1981 ACM Symposium on Symbolic and Algebraic Computation, ACM Inc., New York, 1981, pp. 146–154.

[Strassen, 1969] Strassen,V., Gaussian Elimination is not Optimal. Numer. Math. **13** (1969) pp. 354–356.

[Tarski, 1951] Tarski,A., A Decision Method for Elementary Algebra and Geometry. 2nd ed., Univ. California Press, Berkeley, 1951. MR **10** #499.

[Tournier, 1987] Tournier,E., Solutions formelles d'équations différentielles: Le logiciel de calcul formel DESIR. Thèse d'État, Université I de Grenoble, April 1987.

[Trager, 1976] Trager,B.M., Algebraic Factoring and Rational Function Integration. Proceedings of the 1976 ACM Symposium on Symbolic and Algebraic Computation, ACM Inc., New York, 1976, pp. 219–226. Zbl. 498.12005.

[Trager, 1985] Trager,B.M., On the Integration of Algebraic Functions. Ph.D. Thesis, Dept. of Electrical Engineering & Computer Science, M.I.T., August 1985.

[Viry, 1982] Viry,G., Factorisation des polynômes à plusieurs variables. RAIRO Inform. Théor. **12** (1979) pp. 209–223.

[van der Waerden, 1949] van der Waerden,B.L., Modern Algebra. Frederick Ungar, New York, 1949.

[Wang, 1978] Wang,P.S., An Improved Multivariable Polynomial Factorising Algorithm. Math. Comp. **32** (1978) pp. 1215–1231. Zbl. 388.10035. MR **58** (1979) #27887b.

[Wang, 1980] Wang,P.S., The EEZ-GCD Algorithm. SIGSAM Bulletin **14** (1980) 2 pp. 50–60. Zbl. 445.68026.

[Wang, 1981] Wang,P.S., A *p*-adic Algorithm for Univariate Partial Fractions. Proceedings of the 1981 ACM Symposium on Symbolic and Algebraic Computation, ACM Inc., New York, 1981, pp. 212–217. Zbl. 486.68026

[Wang,1983] Wang,P.S., Early Detection of True Factors in Univariate Polynomial Factorization. Proc. EUROCAL 83 [Springer Lecture Notes in Computer Science 162, Springer-Verlag, Berlin, Heidelberg, New York, 1983], pp. 225–235.

[Wang *et al.*, 1982] Wang,P.S., Guy,M.J.T. & Davenport,J.H., *p*-adic Reconstruction of Rational Numbers. SIGSAM Bulletin **16** (1982) pp. 2–3.

[Wasow, 1965] Wasow,W., Asymptotic Methods for Ordinary Differential Equations. Kreiger Publ. Co., New York, 1965.

[Watanabe, 1976] Watanabe,S., Formula Manipulation Solving Linear ODEs II. Publ. RIMS Kyoto Univ. **11** (1976) pp. 297–337.

[Watanabe, 1981] Watanabe,S., A Technique for Solving Ordinary Differential Equations Using Riemann's *p*-functions. Proceedings of the 1981 ACM Symposium on Symbolic and Algebraic Computation, ACM Inc., New York, 1981, pp. 36–43. Zbl. 493.34002.

[Wilkinson, 1959] Wilkinson,J.H., The Evaluation of the Zeros of Ill-conditioned Polynomials. Num. Math. **1** (1959) pp. 150–166, 167–180.

[Winograd, 1968] Winograd,S., A New Algorithm for Inner Product. IEEE Trans. Computers **C-17** (1968) pp. 693–694.

[Winston & Horn, 1981] Winston,P.H. & Horn,B.K.P., LISP. Addison-Wesley, 1981. (The 2nd ed., 1984, is written for *Common LISP*.)

[Wunderlicht, 1979] Wunderlicht,M.C., A Running-Time Analysis of Brillhart's Continued Fraction Factoring Algorithm. In: Number Theory Carbondale 1979 (ed. M.B. Nathanson) [Springer Lecture Notes in Mathematics 751, Springer-Verlag, Berlin, Heidelberg, New York, 1979], pp. 328–342.

[Yun, 1974] Yun,D.Y.Y., The Hensel Lemma in Algebraic Manipulation. Ph.D. Thesis & Project MAC TR–138, M.I.T., 1974. [reprinted Garland Publishing Co., New York, 1980].

[Yun, 1976] Yun,D.Y.Y., On Square-free Decomposition Algorithms. Proceedings of the 1976 ACM Symposium on Symbolic and Algebraic Computation, ACM Inc., New York, 1976, pp. 26–35. Zbl. 498.13006.

[Yun, 1977] Yun,D.Y.Y., On the Equivalence of Polynomial Gcd and Squarefree Factorization Algorithms. Proc. 1977 MACSYMA Users' Conference NASA publication CP–2012, National Technical Information Service, Springfield, Virginia, pp. 65–70.

[Zassenhaus, 1969] Zassenhaus,H., On Hensel Factorization.J. Number Theory **1** (1969) pp. 291-311. MR **39** (1970) #4120.

[Zippel, 1979] Zippel,R.E., Probabilistic Algorithms for Sparse Polynomials. Proceedings of the 1979 European Symposium on Symbolic and Algebraic Computation [Springer Lecture Notes in Computer Science 72, Springer-Verlag, Berlin, Heidelberg, New York 1979], pp. 216–226.

[Zippel, 1985] Zippel,R.E., Simplification of Expressions Involving Radicals. J. Symbolic Comp. **1** (1985), pp. 189–210.

Index